Programming
in Fortran 90

Programming in Fortran 90

A First Course for Engineers and Scientists

I. M. Smith

University of Manchester, UK

JOHN WILEY & SONS

Chichester · New York · Brisbane · Toronto · Singapore

Reprinted March 1995
Reprinted October 1995
Reprinted August 1999
Reprinted August 2001

Cover illustration The enhanced Mark 1 in June 1949 at the
University of Manchester (© National Computing Centre,
Manchester)

Other Wiley Editorial Offices

John Wiley & Sons, Inc., 605 Third Avenue,
New York, NY 10158-0012, USA

Jacaranda Wiley Ltd, 33 Park Road, Milton,
Queensland 4064, Australia

John Wiley & Sons (Canada) Ltd, 22 Worcester Road,
Rexdale, Ontario M9W 1L1, Canada

John Wiley & Sons (SEA) Pte Ltd, 37 Jalan Pemimipin #05-04,
Block B, Union Industrial Building, Singapore 2057

British Library Cataloguing in Publication Data

A catalogue record for this book is available from the British Library

ISBN 0 471 94185 9

Typeset in 10/12 pt times by Thomson Press (India) Ltd, New Delhi

Contents

Acknowledgement

I gratefully acknowledge the assistance in preparing this book of Dr. D. J. Kidger, who commented on the various drafts and contributed some of the programs in Appendix 5.

The programs in the text are also available on the Internet via anonymous ftp, using the URL `ftp:// golden.eng.man.ac.uk/pub/fe/smith f90.zip`.

1
Introduction

Scientific programming languages—Ada, ALGOL 60, ALGOL 68, APL, PASCAL, PL1 and many others—come and go but the only truly universal language for writing applications programs in engineering and science is FORTRAN. This choice is made on practical rather than intellectual grounds. Even the most recent version of FORTRAN, called Fortran 90, contains features which are thought to be undesirable by some programmers, while lacking other features thought by some to be highly desirable. However the investment by the research and industrial communities in FORTRAN over the past 30 years or so is so vast, and the availability of efficient FORTRAN compilers so widespread, that its continued popularity is inevitable. It is the one language whose availability can be guaranteed on makes of computer ranging from the "PC" to the "supercomputer", ensuring the portability of expensively generated code.

FORTRAN suffers from being an "old" language, and early versions did indeed constrain programmers and encourage a stilted style of program writing. The latest version, Fortran 90, remedies many of these faults and the language is now broadly comparable in the features it provides with, say, the ALGOLs augmented by concepts from the C family.

Brief historical background

The first FORTRANs became available in the 1950s and, although primitive, represented a huge advance on the "machine codes" in which computers had until then to be programmed. Figure 1 illustrates a program in a machine code, actually one of the first programs ever written for a computer capable of storing the program in its own memory, the "Mark I" built at Manchester University in 1948.

The word "FORTRAN" is an acronym for "formula translation" and from the beginning the language gave engineers and scientists, at the time virtually the only users of computers, the ability to encode "formulas" in programs. A root of the quadratic equation $ax^2 + bx + c = 0$ could be found by writing, for example,

```
ROOT=(-B+SQRT(B**2-4.*A*C))/(2.*A)
```

meaning "assign the computed value of the expression after the equals sign, for given values of the variables A, B and C to the variable named ROOT". The advantages of this mode of problem expression over that shown in Figure 1 are indeed staggering.

As the use of programming languages spread, in the late 1950s and early 1960s, the need for standardisation became obvious. Dialects of FORTRAN were developing which meant that programs which ran on one machine would not run on another, i.e. they were not "portable".

In 1966 the American Standards Association (later the American National Standards Institute, ANSI) defined a standard version of FORTRAN, sometimes referred to at the time as FORTRAN IV since it represented the fourth major revision of FORTRAN since its inception. This standard version, the first ever for a programming language, is nowadays known as FORTRAN 66, but its restrictive-

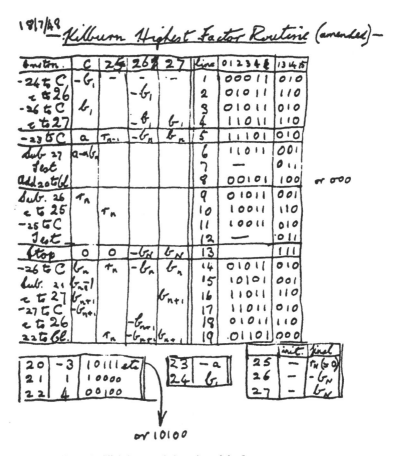

Figure 1. Slightly amended version of the first computer program

ness was such that many implementors did not adhere to the standard, and dialects lacking portability were widespread.

In 1978 a new standard was published, nowadays known as FORTRAN 77. In this version of the language, various improvements were made so that FORTRAN became more compatible with other, more advanced, languages. For example a more "structured" form of programming could be accommodated. However, bearing in mind that one of the great advantages of FORTRAN lies in the preservation of the investment made in "old" programs, most FORTRAN 66 programs were acceptable to FORTRAN 77 compilers. The hope was that programmers would gradually adapt their style to take advantage of the new features provided, but that "old" code should not be obsolete.

The new standard — Fortran 90

In the 1980s developments in FORTRAN continued with a view to creating a new standard to replace FORTRAN 77 in much the same way that FORTRAN 77 had replaced FORTRAN 66. The driving force was the ISO (International Standards Organisation) through a technical committee of the American National Standards Institute (ANSI) called X3J3. The revisions proposed were much more radical and the new FORTRAN was designed to contain many of the features of "advanced" languages such as Ada or ALGOL 68. Originally titled FORTRAN 8x, the new standard was published under the title Fortran 90. As in previous developments, very few features of FORTRAN 77 (strictly speaking none) are rendered obsolete, or unusable, in that they have been deleted from the new version. However various features of FORTRAN 77, which will be detailed later in this book, have been labelled "obsolescent" by the developers of Fortran 90. This means that these features are strong candidates for deletion from any subsequent version of the language, and should not be written into any new code. However "old" codes will still work for the time being.

Perhaps the main new features of interest to engineer/scientist programmers involve the use of arrays. FORTRAN 77 lacked a means for "dynamic" storage allocation which has been remedied in the new version. In addition to leading to much more effective use of storage, passing arrays as parameters in SUBROUTINEs has become much simpler. Extra features include the definition of various operations, such as addition, to be valid for arrays, and the inclusion of useful array operations such as multiplication and transposition as intrinsic procedures. This should allow optimisation of array calculations on various "vector" and "parallel" machine architectures.

Scope of this book

The book is intended to serve as a "mini-manual" illustrating the main features of Fortran 90 likely to be of use in typical engineering/scientific programs. Emphasis is given to array handling and subroutine structures since these are such strong features of applications programs. The material is initially arranged in short chapters each of which might form the basis of a single lecture in an introductory Fortran 90 course. Illustrative pieces of program are given in every chapter, along with further exercises for students to attempt on their own with model answers in an Appendix. Differences from FORTRAN 77 and obsolescent features are described as they arise.

Later chapters deal with less widely used features of the language and Appendices are devoted to lists of the FORTRAN character set, intrinsic procedures and I/O options and a glossary of terms used.

No attempt is made to describe every feature of what is now a very large language in detail. Interested readers are referred to *Fortran 90 Explained* by M. Metcalf and J. Reid, Oxford Science Publications (1990), or to *Fortran 90 : A Programmers' Guide* by W. S. Brainerd, C. H. Goldberg and J. C. Adams, McGraw-Hill (1990).

2

Getting Going: Names, TYPES, Simple Input/Output and Program Structure

There are names and there are things. A name is a spoken sound which designates a thing and acts as a sign for it

de Montaigne

Introduction

In the previous chapter, a Fortran 90 statement was given for finding a root of a quadratic equation. This took the form

```
ROOT = ( −B + SQRT (B**2 − 4.*A*C) ) / (2.*A)
```

In this statement, A, B, C and ROOT are names of variables. The symbols + and − have their usual algebraic meaning while * represents multiplication, / division and ** exponentiation, or raising to a power. Parentheses are used to collect terms in expressions and SQRT takes the square root of the expression in parentheses after it. The symbol =, however, does *not* mean "equals" but rather "takes the value" or "becomes". What the statement signifies is that the variable ROOT takes the computed value of the expression after the = sign. For completeness the full Fortran 90 "character set" of symbols is listed in Appendix 1.

As in any formal language, names of variables in Fortran 90 (also sometimes called "identifiers") must conform to certain fixed rules. They must be composed of the alphanumeric characters (the letters of the English alphabet and the Arabic numerals 0 to 9) supplemented by the underscore symbol, _. The maximum number of characters in a name is 31 and the first character must be a letter. There are no reserved words but it is inadvisable to use for example SQRT as a variable name since confusion is likely. The underscore symbol is usually used to separate different parts of a name, since a space or blank is not a valid character in a name. Thus

```
CHICAGO_BEARS
MANCHESTER_UNITED
HEART_OF_MIDLOTHIAN
TRASH
XYZ_123
Z
```

are valid names. Although both lower and upper case letters can be used for names, for clarity upper case has been used in this book; readers are advised to use all upper case or all lower case throughout a program unit. Note that FRED, Fred, freD etc are all equivalent. This is the first time that mixed character modes have been allowed in the standard and we have implicitly recognised it already by calling the language Fortran 90 as distinct from earlier FORTRANs. The convention adopted in the book is that Fortran 90 names, and the syntax of the language, are given in upper case, but that "comments" and output text, which are not strictly part of Fortran 90 are given in lower or mixed case.

Examples of invalid names are

49ERS	Does not begin with a letter
MAN.UTD	Full stop not valid character
L A RAMS	Space (blank) not valid character
A/B	Divide symbol/not valid character

Simple variable types

We shall see in due course that variable names can refer to different "types" of quantities, for example to non-numeric characters. The simplest of these TYPEs are single INTEGER and REAL numbers. The former are usually reserved for counting, for example when some operation is to be carried out a whole number of times and the latter for arithmetic calculations.

In most other languages it is necessary for the programmer to "declare" the names of variables to be used later in a program at the beginning of the program together with their TYPE. This kind of declaration statement is allowed in Fortran 90, so that one can write

```
REAL : : ALPHA, BETA, GAMMA, DELTA, EPSILON, IOTA
INTEGER : : A1, A2, A3
```

as typical declarations (but be careful: EPSILON is like SQRT and used for other purposes). However, Fortran 90 also allows "implicit" typing which means that any name beginning with letters in the range I to N is assumed without declaration to refer to an INTEGER variable while names beginning with any other letter are assumed to refer to REAL variables, unless declared otherwise.

It is good practice to declare all the variables used at the beginning of programs but many experienced FORTRAN programmers do not. Because the "implicit" TYPE convention is so deep-rooted, programs may be more easily understood if I, J etc are used as counters and subscripts.

To be completely safe, it is possible to turn off implicit typing altogether by the statement

```
IMPLICIT NONE
```

at the beginning of a program. All variable names must then be declared. This is good practice because some programming errors can be detected in this way, for example misspelt variable names or inconsistent use of variable names.

We shall see in due course what happens when REAL quantities are assigned to INTEGER variables and vice-versa.

Simple input and output

Nowadays programmers usually communicate with their programs by means of a keyboard and VDU screen. Information can be fed into the program, or "input", by typing it on the keyboard while information can be retrieved from the program by its display, or "output", on the screen.

In order to make a start with Fortran 90, we introduce here a very simple form of input and output called "list-directed" input/output. For the program to receive input from the keyboard we can use a statement of the following form:

```
READ *, ONE, TWO, THREE
```

This statement would allow three (REAL) numbers to be typed, one after the other, on the keyboard and they would be assigned the variable names ONE, TWO and THREE respectively when the "ENTER" key is struck. Obviously there is a need to be able to separate the numbers from each other and this can be done by typing one or more blanks (spaces), a comma, optionally preceded and optionally followed by blanks, or a slash (/) optionally preceded and optionally followed by blanks. Thus typing

```
3.5   4.7  8.2      (ENTER)
3.5,4.7,8.2         (ENTER)
3.5/ 4.7/ 8.2       (ENTER)
```

or even

```
3.5     (ENTER)
4.7     (ENTER)
8.2     (ENTER)
```

would all lead to ONE having the value 3.5, TWO the value 4.7 and THREE the value 8.2 (or as close to these values as the particular computer can achieve).

The complementary statement allowing variables to be displayed on the screen takes the form:

```
PRINT *, ONE, TWO, THREE
```

In "list-directed" input and output, the asterisk implies reading from the keyboard or writing to the screen in "free format" (see later) and does not form part of the special Fortran 90 words READ and PRINT which are examples of "keywords".

Program layout

Fortran 90 programs have to adhere to a specified layout or "source form" to be successfully understood by the "compiler" which is a program in the computer which translates FORTRAN into machine code. In the early FORTRANs the program layout was very rigidly specified and linked to the appearance of information on punched cards. Thus, for example, statements used not to be allowed to appear before the seventh "column" on the punched card, usually necessitating the typing of six blanks at the beginning of every line. Fortunately these restrictions have been abolished in Fortran 90 and statements can be typed at will, with few restrictions.

The VDU screen on which a program appears allows statements to be written on lines, at most 132, but typically 80 or 120, characters long.

The end of a program must be signified by the statement END and it is strongly advised that the start of a program be signified by the statement PROGRAM followed by a blank or blanks and a name for that particular program (following the above rules for Fortran 90 names). The same name and the word PROGRAM can be used in the END statement after a blank or blanks. Thus a typical program layout is

PROGRAM FIRST

(Various Fortran 90 statements)

END PROGRAM FIRST

The special characters ! & and ;

It will be obvious, even from the above specimen program, that some care is needed in the use of the blank character. In many senses blanks are ignored and can be used to improve layout, but in other instances blanks are essential as separators, for example between PROGRAM and FIRST.

Completely blank, and uncommented, lines are allowed.

Good programs are always annotated by the writer with information about what the program is trying to achieve. The method of annotation or "commenting" allowed in Fortran 90 is that all characters on a line following an exclamation mark (!) are ignored. Therefore whole lines can be used for descriptive purposes or a single statement can be "commented" by following it by ! and text.

Although it is seldom necessary in good programming practice, statements can sometimes exceed the length of a line allowed at a particular installation (typically 80 or 120 characters). If this should occur, the ampersand character (&), usually placed at the end of a line, allows continuation of the statement onto the next line. In fact it is safer to place another ampersand at the beginning of each continuation line as well. The maximum length of a statement is 2640 characters, including blanks, or 40 lines, but implementations differ.

With typically 100 characters per line available for statements, it is clearly wasteful that short statements should take up a full line. Fortran 90 therefore allows multiple statements to occur on one line, separated by the semi-colon character (;).

Normally, statements terminate at the end of a line where the ; terminator is not necessary.

A very simple program to get going

The following program is extremely simple but illustrates some features of Fortran 90 and of programming in general. It merely reads some numbers typed in from the keyboard and prints them on the screen in a different order. Actually this is an example of "echoing" input data which is good practice (at least for small amounts of data).

```
    PROGRAM VERY_SIMPLE
! this program reads 6 real numbers in sets of 3
! and prints out the sets in reverse order
    IMPLICIT NONE
REAL : : ONE, TWO, THREE, &
& FOUR, FIVE, SIX
    READ *, ONE, TWO, THREE; READ *, FOUR, FIVE, SIX
    PRINT *, SIX, FIVE, FOUR; PRINT *, THREE, TWO, ONE
    STOP
      END PROGRAM VERY_SIMPLE
```

The only statement in this program we have not seen before is STOP which causes termination of the program. It can be used more than once in any program but its most traditional use has been once preceding the END statement as in VERY_ SIMPLE. However it is really redundant in this position and our convention will be to omit it.

Note again that we have adopted a convention that comments, which are not "executable statements" in Fortran 90, are typed in lower case.

On a typical installation, when a program like the above is compiled and a "run" command issued, a "prompt" symbol, such as a query (?), may appear on the screen. This invites the programmer to key in numbers associated with the list-directed READ statements in the program. One computer used by the author has a "default wordlength" for storing real numbers of 32 binary "bits". When the appearance of the screen after keying in some real numbers was as follows:

```
?
8.5, 4.4, 3.7     (ENTER)
?
9.8, 6.6, 1.9     (ENTER)
```

the output produced on the screen by the list-directed PRINT statements in the program was as follows:

```
1.89999962   6.60000038   9.80000019
3.69999981   4.39999962   8.50000000
```

Using the same computer, it is possible to increase the "wordlength" for storing real numbers to 64 binary bits. When this is done, and precisely the same input is keyed in, the result on the screen becomes:

1.89999999999999999991   6.60000000000000000009   9.80000000000000000004
3.69999999999999999996   4.39999999999999999991   8.50000000000000000000

The program has clearly worked "correctly", but in the sense that 6.60000038 or 6.60000000000000009 are accurate enough representations of 6.6 for practical purposes. Clearly the two representations of 6.6 are not equal to 6.6 or to one another, and later we shall have to be careful about testing REAL numbers to see whether they are "equal".

If it is attempted to input all 6 numbers to VERY_SIMPLE by a single line on the keyboard, i.e.

?
8.5/4.4/3.7/9.8/6.6/1.9   (ENTER)

the result will be another prompt (?) on the screen. Each call to the list-directed READ statement, and there are two of these, separated by a semi-colon, reads information from a new "record" or line. Thus the intended input numbers 9.8, 6.6 and 1.9 would be lost and would have to be input again in the form of a new record, as in our first run of the program.

It is possible to economise if series of identical input values occur. For example if three consecutive input numbers had the value 8.5 we could type 3*8.5.

On some systems no "prompt" symbol appears on the screen and when a "run" command is issued, "Program entered" or some similar phrase may appear, inviting numbers to be typed in. Best practice is for the program itself to prompt the user to enter the appropriate numbers. Examples are given in subsequent chapters.

Differences from FORTRAN 77

Even in this short introductory chapter, several radical differences between Fortran 90 and FORTRAN 77 are apparent. These relate to names, and to program layout. In FORTRAN 77, names were restricted to six characters and the underscore was not included. This had an inhibiting effect on program readability, by rendering many natural names invalid.

There was no double colon symbol in FORTRAN 77 and so declarations took the shortened form such as

```
REAL VAR1, VAR2, VAR3
```

which is still legal since all of FORTRAN 77 is contained within Fortran 90.

Program layout in FORTRAN 77 still reflected the "field format" of punched card images, leading to a cramped and unnatural style. The screen was divided into 80 (card) columns and commenting was achieved by putting the character 'C' in column 1. Columns 1 to 5 were usually used for numerical "labels" or pointers to statements which we shall see have almost disappeared from Fortran 90 programs. Statements could be continued over more than one line by placing a continuation symbol in column 6. Program statements could only occupy columns 7 to 72. Multiple statements on a line were not allowed.

To enable FORTRAN 77 style code to be recognised by a Fortran 90 compiler, a FIXED source form (in contrast to the new FREE one) is allowed.

Exercises

1. Which of the following are not valid Fortran 90 names?

   (a) POLTERGEIST
   (b) 7UP
   (c) R2D2
   (d) PH.D.
   (e) A/B/C
   (f) SQRT
   (g) B SC
   (h) F(X)

2. What is wrong with the following program? Apart from that, can its style be improved?

```
PROGRAM TEST
    READ*,A,B,C
    PRINT*,A,B,C,D
    STOP
END
```

Solutions are given in Appendix 5.

3
Simple Arithmetic

Introduction

We have already seen that arithmetic can be done in Fortran 90 in terms of two TYPEs of variable called INTEGER and REAL. To these we can add a third TYPE called COMPLEX which enables arithmetic operations to be carried out on complex numbers represented in the usual form a + ib. The two components (real part and imaginary part) of the complex number are contained in parentheses and separated by a comma. Thus the declaration

```
COMPLEX :: C
```

could be followed by the assignment for constant real and imaginary parts

```
C = (1.0, 2.0)
```

leading to C taking the value 1.0 + 2.0i. Assignments of variables to complex numbers involve the intrinsic procedure CMPLX (see Chapter 4).

Simple TYPE declaration statements

All the different varieties of declaration statements in Fortran 90 *must* appear at the beginning of programs, before any executable statements (such as the READ, PRINT and assignment statements we have already seen). In due course we shall see that there are many forms of "TDS" and that programmers can create their own data TYPEs but we shall concentrate at present on simple TDSs involving the standard numerical TYPEs namely REAL, INTEGER and COMPLEX. The other two "intrinsic" TYPEs which will be introduced in due course are LOGICAL and CHARACTER.

Our usual requirement is to be able to declare the names of variables, whose values change during the course of a program's execution. However, a few names will usually represent constants, whose values we wish to protect so that they remain unchanged during program execution.

Declaration of variables follows the rules we have already seen, typically

```
REAL : : VAR1,VAR2,VAR3
INTEGER : : I1,I2,I3
COMPLEX : : C1,C2,C3
```

In the case of constants, the word PARAMETER is included in the declaration, and protects these names during program execution. However such constants have also to be "initialised" at the time of declaration, for example

```
INTEGER,PARAMETER : : HOURS_IN_A_DAY = 24
REAL,PARAMETER : : PI = 3.1415926536
```

If it seems appropriate, variables can also be initialised in the declaration statement, for example

```
INTEGER : : DAYS_IN_A_MONTH = 31
```

Representation of numbers

In the previous chapter we met an immediate difficulty in the representation of REAL numbers. Depending on the precision with which a given computer can store data the real number 5.7 can be represented in a variety of forms which are very close to, but usually not exactly equal to, the decimal number 5.7.

In the case of INTEGERs, there is no equivalent difficulty in that -256 or 1 or $+25$ would be represented in the same unambiguous way in any practical computer. The '+' sign is optional for positive integers.

For INTEGERs, therefore, the only question to be answered is the *range* of INTEGER numbers which a given processor can support. We shall see later in Chapter 11 that "intrinsic functions" are provided in the language which enable this range to be found for a given computer and indeed changed. For the moment, however, it is only necessary for us to know that the range of INTEGERs which can be represented on a typical computer is very large, and adequate for most practical purposes. For example for a computer with an internal wordlength of 16 binary bits (shown in the previous chapter to be capable of representing REAL data with less than six decimal place precision) the range of INTEGERs which can be accommodated is typically from -32768 to $+32767$.

REAL numbers

These are represented more generally in "floating point" form, of which an example is

$$+56.25E-5$$

meaning the REAL number $+0.0005625$, or as close to it as the processor can get. This full representation consists of an (optionally) signed integer part ($+56$), a decimal point (.) a fractional part (25) and an optionally signed "exponent" part

(E–5). Only one or both of the integer part and the fractional part are essential and it is advisable although not essential always to include the decimal point although the exponent part can be absent.

Typical REAL numbers can be represented by the following:

3.14159
1.
− .5
+ 26.85
− 1.E5

Fortran 90 arithmetic operators

The full list of these operators is given below:

**	exponentiation (raising to a power)
*	multiplication
/	division
+	addition
−	subtraction

These operators are listed in decreasing order of "binding strength" with dashed lines separating groups of equal binding strength. Thus

A*B−C+D

is evaluated as A*B from which C is then subtracted and finally D added. Precedence is therefore left to right amongst operators of equal strength. The exception to this rule is exponentiation, where 2**3**2 is 2**9 or 512.

The natural order can be overridden by parentheses or by surplus operators. For example the quantity A.B/C.D can be represented by (A*B)/(C*D) or by A*B/C/D. Other examples are 2*3**2 which gives 18 and 4−2+2 which gives 4.

Operators may not appear consecutively; for example 2** −1 is not allowed, and would have to be written 2**(−1) or in some other equivalent form.

Raising to a power

A word of caution is justified when the exponentiation operator ** is used to raise quantities to a non-integer power. Integer exponentiation causes no difficulty because 2**3 can just be recognised by the compiler as 2*2*2. However A**B can be written where A and/or B can be REAL, INTEGER or COMPLEX, and will in general be evaluated as $e^{B \log A}$. Hence if A is negative REAL, it cannot be raised to a REAL power. Care is necessary since, for example, $(-1) **0.5** (-1)$ is 1.0 whereas $-1**0.5**(-1)$ is -1.0.

The arithmetic assignment statement

We have already met an example of such a statement in Chapter 2. The general form is:

    variable = expression

and means "evaluate 'expression' and assign its value to 'variable' ". We repeat that "=" does *not* signify "equals".

Typical assignment statements might be

```
A = A + B                ! = does not mean ''equals''
PI = 3.14159             ! an approximation to pi
VOL = PI*R*R*H           ! assuming pi given by previous statement
AREA = DEPTH*WIDTH
TRIANGLE_AREA = 0.5*BASE*HEIGHT
```

A "TYPE conversion" is necessary if the TYPE of 'variable' is different from the TYPE of 'expression'. The TYPE of an arithmetic expression depends on the TYPEs of its operands. Thus, the result of an operation is of the same TYPE as the operands if both are of the same TYPE (4/2 is of TYPE INTEGER), while if the operands are of different TYPEs, then the one of TYPE INTEGER is first converted to TYPE REAL and the result will be of TYPE REAL (3.6/2 is converted to 3.6/2.0 which is 1.8).

Whatever the TYPE of the value of 'expression' it has to be converted to the same TYPE as 'variable' before it can be assigned to 'variable'.

Truncation

In many computers, INTEGER quantities take up less space than REALs, but in any event a conflict always arises when REAL quantities are assigned to INTEGER variables. The general rule is that a REAL is "truncated", that is, stripped of its decimal fractional part, before assignment to an INTEGER variable. Thus

```
I = 7.6
```

results in I taking the value 7. If the REAL variable R has the value 3.6, the arithmetic assignment statement

```
I = 2*R/4
```

results in I taking the value 1. Other examples are

```
I = 2** (-3)
```

resulting in I taking the value 0, while the results of $6/3$, $8/3$ and $-8/3$ are $2, 2$ and -2 respectively.

There is no significant conflict in INTEGER to REAL conversions in that 2 becomes 2.0, or very near it, and so on, but a general recommendation is that these "mixed mode" expressions should be avoided wherever possible. Programmers should realise clearly the TYPEs of variables intended and use them appropriately.

Truncation can be used to advantage in that the expression

```
J/5*5
```

will take the values 0,0,0,0,0,1,1,1,1,1,2,2 etc as J takes the values 0,1,2,3,4,5,6,7,8,9, 10,11 etc and could be used to discriminate on every 5th J in an ascending INTEGER sequence.

Printing out text

In Chapter 7 we shall see that text can be stored in special variables called "character" variables, and thus manipulated as input or output. For the moment,we want to make output from our simple programs more intelligible using the "list directed" output statement described in Chapter 2. We can combine output of text with output of numerical values of variables by including the text in single or double quotes in the PRINT statement. For example

```
PRINT*, "The Value of Variable 1 is", VARIABLE_ONE
```

Since output text is not "executable" Fortran 90, we allow mixing of upper and lower case as in this example.

Appendix 5 gives many more examples as answers to exercises.

Differences from FORTRAN 77

Apart from the form of the TYPE declaration statement, there are no significant variations from FORTRAN 77 in the material described in this Chapter, which has been deliberately kept at a simple level. However we shall see in due course that Fortran 90 gives much greater control over the precision with which REAL and INTEGER variables are represented via the KIND = and LEN = specifiers. In FORTRAN 77, REAL variables could only be expanded by the DOUBLE PRECISION declaration, which is a candidate for obsolescence.

As an alternative to initialising variables in a TYPE declaration statement as described above, there exists in FORTRAN 77 the DATA statement. For example

```
DATA IDEE, IBEE, IDBEE, IH, IFLOW/5*4/
```

would initialise five INTEGER variables with the value 4. Since initialisation can always be achieved in other ways in Fortran 90, the DATA statement is also a candidate for obsolescence.

As mentioned in the previous chapter there is no double colon symbol in FORTRAN 77 and declarations can only take the typical form

```
REAL A, VOL, AREA
```

Consequently declaration with initialisation is not possible. The PARAMETER statement for initialisation of constants takes the form

```
PARAMETER(IA=200, IB=300)
```

in FORTRAN 77.

The option of single or double quotes to encapsulate text is new. Previously only single quotes were allowed.

Exercises

1. Assuming variables of implicit TYPE, suggest what the results of the following apparently identical pieces of program would be:

```
P=5
Q=-9
R=4
X=P*Q/R
Y=Q/R*P
Z=P/R*Q
```

and

```
I=5
J=-9
K=4
X=I*J/K
Y=J/K*I
Z=I/K*J
```

Test your suggestions by writing Fortran 90 programs to carry out the arithmetic.

2. The formula for converting temperatures in degrees Centrigrade (°C) to degrees Fahrenheit (°F) is:

$$°F = °C*(9/5) + 32°$$

Suggest how this conversion would be done by the following Fortran 90 statements assuming implicit TYPE:

```
FAHRENHEIT=1.8*CENTIGRADE+32.0
FAHRENHEIT=9*CENTIGRADE/5+32
FAHRENHEIT=9/5*CENTIGRADE+32
```

and test your suggestions by programming.

3. Assuming variables of implicit TYPE, what values do X,Y,M and N have after this program extract?

```
A=17.5
B=A+3
I=B-6.6
M=I-9*2
X=2*A+M
I=X/I
Y=A+I/M
N=A+3*I/M
```

Test by programming the extract.

4. Write Fortran 90 expressions corresponding to the following:

(a) $x + y^3$

(b) $(x + y)^3$

(c) x^4

(d) $\dfrac{a+b}{c}$

(e) $\dfrac{a+b}{c+d}$

(f) $\dfrac{a+b}{c+\dfrac{d}{e+f}}$

(g) $1 + x + \dfrac{x^2}{2!} + \dfrac{x^3}{3!}$

5. Which of the following are valid Fortran 90 numbers?

(a) -85 (b) $+3E+2$ (c) $20{,}000$ (d) $-.7E6$
(e) $50*E-7$ (f) $+90\,653\,876$ (g) ±4.37

Test by reading them into the computer and printing them back out again.

6. Consider a general triangle with sides A,B and C, and its inscribed and escribed circles (radii IR and ER respectively) as shown below.

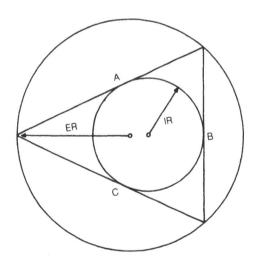

The respective radii are given by

$$IR = \sqrt{\dfrac{(S-A)(S-B)(S-C)}{S}}$$

$$ER = \dfrac{ABC}{4\sqrt{S(S-A)(S-B)(S-C)}}$$

where $2S = A + B + C$.

Write a program to read in the three sides and calculate and print the radii of the inscribed and escribed circles. Check your answers by completely independent means, for example by pocket calculator, for the cases $A = B = C = 1.0$ and $A = 65.23$, $B = 39.77$, $C = 58.05$.

Note: You may assume that SQRT (expression) can be used to calculate a square root. Write out a plan of the structure of your program, in symbols and in words, before doing any coding.

4

Simple Intrinsic Procedures

Introduction

In due course we shall see that large programs consist of a collection of smaller pieces of program or "subprograms" which carry out pre-defined tasks such as solving a set of linear simultaneous equations. There are two kinds of subprograms available in Fortran 90, called FUNCTIONs and SUBROUTINEs respectively. The essential difference between these is that the output from a FUNCTION is a single quantity (even although it may be an "array"—see Chapter 8) whereas the output from a SUBROUTINE can consist of several quantities of various TYPEs. The tasks carried out by some subprograms, especially FUNCTIONs, are so widely used in applications programs in engineering and science that they are provided as part of the language. These subprograms are called "intrinsic procedures".

Intrinsic FUNCTIONS

We have already seen one such FUNCTION in the previous chapters. In solving a quadratic equation we had to evaluate the quantity $\sqrt{b^2 - 4ac}$. In order to do this in FORTRAN we wrote an expression SQRT (B*B − 4.*A*C) where SQRT is an example of an intrinsic FUNCTION. The "argument" of the FUNCTION is contained in the parentheses and by including SQRT (----) in a program we say that we "call" the FUNCTION with its particular argument (a REAL number in this case). We also speak of the output or result of having called the FUNCTION being "returned" to the program which called it.

A short list of intrinsic FUNCTIONS

There are over 100 intrinsic procedures available in Fortran 90 which are listed in full in Appendix 3. The list given below is restricted to a few of the more commonly used procedures which beginners are likely to need early in their programming experience. A subdivision has been made into FUNCTIONs which convert their arguments, those which test them, and those which perform basic mathematical operations on them.

(i) Conversion

Task	Usual definition	FUNCTION name	TYPE of argument	TYPE of result
Absolute value	|x|	ABS(X)	REAL	REAL
Truncation	Integer part of x	INT(X)	REAL	INTEGER
Nearest	Nearest integer to x	NINT(X)	REAL	INTEGER
	Nearest integer not less than x	CEILING(X)	REAL	INTEGER
	Nearest integer not greater than x	FLOOR(X)	REAL	INTEGER
Conversion to REAL		REAL(I)	INTEGER	REAL

(ii) Testing

Task	Usual definition	FUNCTION name	TYPE of arguments	TYPE of result
Remainder	Remainder of x/y	MOD(X,Y)	INTEGER REAL	INTEGER REAL
		MODULO(X,Y)	INTEGER REAL	INTEGER REAL
Choose largest value	max(x,y,z,...)	MAX(X,Y,Z)	INTEGER REAL	INTEGER REAL
Choose smallest value	min(x,y,z,...)	MIN(X,Y,Z)	INTEGER REAL	INTEGER REAL

Some explanation of the remaindering FUNCTIONs called MOD and MODULO may be necessary. Whereas MOD(X,Y) is just X-Y* INT(X/Y), MODULO(X,Y) is X-Y* FLOOR(X/Y). They differ when negative arguments are used. For example MOD(3.5, -2.0) is 1.5 whereas MODULO (3.5, -2.0) is −0.5.

(iii) Mathematical operations

Task	Usual definition	FUNCTION name	Usual TYPE of argument	TYPE of result
Square root	$\sqrt{x}$	SQRT(X)	REAL	REAL
Exponentiation	exp(x)	EXP(X)	REAL	REAL
Natural logarithm	$\log_e x$ or $\ln x$	LOG(X)	REAL	REAL
Common logarithm	$\log_{10} x$	LOG10(X)	REAL	REAL
Sine of an angle	sin x (radians)	SIN(X)	REAL	REAL
Cosine of an angle	cos x (radians)	COS(X)	REAL	REAL
Tangent of an angle	tan x (radians)	TAN(X)	REAL	REAL
Hyperbolic sine	sinh x (radians)	SINH(X)	REAL	REAL
Hyperbolic cosine	cosh x (radians)	COSH(X)	REAL	REAL
Hyperbolic tangent	tanh x (radians)	TANH(X)	REAL	REAL
Arcsine	$\sin^{-1} x$ (principal value)	ASIN(X)	REAL	REAL
Arccosine	$\cos^{-1} x$ (principal value)	ACOS(X)	REAL	REAL
Arctangent	$\tan^{-1} x$	ATAN(X)	REAL	REAL

(iv) Complex arithmetic

Task	FUNCTION name	TYPE of arguments	TYPE of result
Create COMPLEX number from "real" and "imaginary" variable components X and Y	CMPLX(X,Y)	REAL or INTEGER	COMPLEX
Find COMPLEX conjugate	CONJG(Z)	COMPLEX	COMPLEX

Although mathematical operations are usually carried out on REAL numbers, COMPLEX arguments can be used with SQRT, SIN, COS, EXP and LOG.

Calculations using complex arithmetic

Typical operations in complex arithmetic are therefore easily accomplished in Fortran 90. For example

$$Z = \sqrt{(5 - i)\log(3 + 2i)}$$

$$Z = \sin(2 + 3i)$$

$$Z = \left[(2 + 3i)(1 - i) - \frac{3e^{5i}}{\sqrt{2 - i}} \right]^{(1/2 - 2i)}$$

could be computed using the small program:

```
PROGRAM COMPLEX_NUMBERS
! examples from a text on C++
IMPLICIT NONE
COMPLEX :: A, B
  A = (5.0, -1.0); B = (3.0, 2.0)
  PRINT*, SQRT (A*LOG(B))
  PRINT*, SIN ((2.0, 3.0))
  A = (2.0, 3.0)*(1.0, -1.0)
  B = 3.0*EXP((0.0, 5.0))/SQRT ((2.0, -1.0))
  PRINT*, (A - B)**(0.5, -2.0)
END PROGRAM COMPLEX_NUMBERS
```

The result is:

(2.6640487, 0.3110938)
(9.1544991, −4.1689072)
(−7.0289340, −2.0416074)

Differences from FORTRAN 77

The standard FUNCTIONs for taking logarithms tended to use "specific names" such as ALOG(X) and ALOG10(X) in FORTRAN 77, depending on the TYPE of the argument, rather than the "generic names" given above. Other examples of such "specific names" in FORTRAN 77 are AMOD (for REAL arguments), MAX0 (for INTEGER arguments), AMAX1 (for REAL arguments), MIN0 (for INTEGER arguments) and AMIN1 (for REAL arguments). When we come to later examples we shall see that a major change is that a FUNCTION can return an array in Fortran 90 whereas it could not in FORTRAN 77.

Exercises

1. It is required to find the vertical deflection of an elastic beam bearing on an elastic foundation. The foundation has a stiffness, k, of $4*10^3$ kN/m² and the length of the beam, L, is 10 m. The flexural rigidity of the beam, EI, is 10^6 kN/m². For a central vertical load P, the vertical deflection v at distance x from the load is given by

$$v = \frac{2P\beta[\sinh(\beta L)\cos(\beta x)\cosh(\beta(L-x)) - \sin(\beta L)\cosh(\beta x)\cos(\beta(L-x))]}{k[\sinh^2(\beta L) - \sin^2(\beta L)]}$$

where

$$\beta = \sqrt[4]{\frac{k}{4EI}}$$

Find the deflection at distances 0, 2 and 5 m from a 100 kN load.

2. To a first approximation a satellite may be assumed to move in a circular orbit round the Earth. The velocity V_c in m/sec necessary to maintain the orbit of a satellite at

a distance of h m above the surface of the Earth is approximately

$$V_c = \frac{7749\sqrt{R}}{\sqrt{R+h}}$$

where R is the radius of the Earth (6.2712×10^6 m). The escape velocity, V_e, at any particular altitude is $V_e = \sqrt{2}V_c$. Write a program to read in an altitude h, compute the circular velocity and the escape velocity (in km/hour) and the time taken to complete one orbit (in minutes).

Test your program for heights of 500 m and 15,000 m.

Hint: Intrinsic FUNCTIONs can be used to calculate π, for example as `4.0*ATAN(1.0)` or `ACOS(-1.0)`.

3. A formula for the cosine of an angle x is $\cos(x) = (e^{ix} + e^{-ix})/2$. Use the Fortran 90 intrinsic FUNCTIONs called CMPLX and EXP to find the cosine of $\pi/4$.

5
Repetition

Introduction

One of the most essential features in a programming language is the capacity for a statement, or more probably a series of statements, to be executed more than once. For example we may wish to find the roots of several quadratic equations $ax^2 + bx + c = 0$ which differ only in their coefficients a, b and c. The programming structure which enables this to be achieved is called a "loop" and since various things are to be "done" repetitively, the structure is called in Fortran 90 a "DO loop".

Structure charts

As programs become more complicated, it will prove to be useful to have a pictorial way of describing their structure in much the same way as engineering drawings are used to describe a complicated piece of machinery or a building. The pictorial forms we shall describe are called "structure charts" and the representation of a repetitive structure or loop which we will use is shown below:

Number of repetitions
Things to be done

We shall see that a common program structure involves repetitions of repetitions and so we expect a "nested" arrangement of loops to appear as follows:

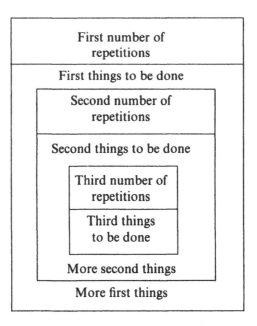

The DO statement with counter

A typical construct might be:

```
ITERATIONS:    DO I = 0,8,2
                    (Various statements)
               END DO ITERATIONS
```

in which "ITERATIONS" is the name of the loop which can be used informatively to supplement comments. If a name is given to a loop before the colon, the same name must appear in the END DO statement. However, the naming of loops is optional, and

```
DO I = 0,8,2
    (Various statements)
END DO
```

is equally valid. The INTEGER I is called the "control variable" or counter since it controls or counts the number of times the loop will be executed. In the above example it takes the value 0 and is then incremented in steps of 2 until it reaches 8. Thus the statements enclosed by the loop would be executed five times in this case. In some loops control variables like I have no other purpose but to keep a count of the number of iterations and do not appear in any of the statements within the loop. In other instances, control variables can appear also in the internal loop statements. If the increment (2 in the above example) is missing, an increment of 1 is implied.

A loop to read in N REAL numbers and print their sum might be:

```
SUM = 0.0      ! initialise sum
         READ*, N
SUMMATION:    DO I = 1, N
                READ*, A
                SUM = SUM + A
              END DO SUMMATION
              PRINT*, SUM
```

Note that the variable SUM has been "initialised" by setting it to zero before entering the loop. One cannot assume the value of any declared or used variable on entry to a program.

A loop to calculate and print factorial N might be:

```
FACTORIAL_N = 1      ! initialise factorial N
         READ*, N
FACTORIAL:        DO I = 1, N
                    FACTORIAL_N = FACTORIAL_N*I
                  END DO FACTORIAL
                  PRINT*, FACTORIAL_N
```

In the first example the control variable is never used for another purpose whereas in the second it is.

The lower and upper limits of the control variable (0 and 8 in our first example) and the increment (2 in the first example) can in general be any INTEGER expression, positive, negative or zero. Thus

```
DO IP = NYE*NXE, 0, -NZE
```

is legal but the difficulty arises that a whole number of subtractions of NZE from NYE * NXE may not yield zero. Therefore there is a rule for the (INTEGER) number of times a loop is obeyed. If in general we write

```
DO VARIABLE = EXP1, EXP2, EXP3
```

the loop will be executed a number of times given by the INTEGER part of (EXP2−EXP1 + EXP3) /EXP3. If this quantity is negative or zero, the loop will not be executed at all.

The language defines the value of the control variable on exit from a DO loop by performing the increment at the end of the loop whether or not the iteration is the final one. Thus in our earlier example

```
DO I = 0,8,2
  (Various statements)
END DO
```

the variable I would have the value 10 on completion of the loop. Utilisation of this property does however lead to obscure programs and it is strongly recommended that control variables be re-set after leaving a loop.

Endless loops

A very simple and useful version of the loop which does not require a control variable is provided, and can be left by issuing an EXIT or CYCLE statement. The forms are shown in parallel below:

```
ENDLESS: DO          REPEAT: DO
  (Various statements)   (Various statements)
     EXIT ENDLESS        CYCLE REPEAT
  (Various statements)   (Various statements)
     END DO ENDLESS      END DO REPEAT
```

In the "ENDLESS" version, EXIT causes the next statement *after* the loop to be executed whereas in the CYCLE version, if there are more iterations to be done, they will be initiated until END DO is exhausted. That is, EXIT is a terminating statement, essentially transferring control to the end of the loop, while CYCLE transfers it back to the beginning, having terminated the current iteration only.

The control over the EXIT or CYCLE processes, which can be used in loops of any type, would normally take the form of a conditional statement. These are described in Chapter 6 which follows, and so a full description of endless loops, with examples, must be deferred till then.

Differences from FORTRAN 77

In Chapter 2 the program layout demanded by FORTRAN 77 was described with its rigid 80-column "punched card" image format. Numerical "labels" could be attached to statements by typing them to the left of the statements, in "columns" 1 to 5. This was necessary because the FORTRAN 77 DO statement contained a label and took the typical form:

```
DO 100 I = 0,8,2
(Various statements)

100 CONTINUE
```

Thus the DO loop terminated at the statement labelled 100, which in the above example is the "do nothing" statement, CONTINUE. It was even possible to end a loop at an executable statement and to terminate nested DO loops at the same labelled statement but this is now considered obsolescent (see below).

The abolition of the label in the DO statement leads to profound improvements in program layout. In FORTRAN 77 it was perfectly possible to "indent" programs (within the confines of "columns" 7 to 72) but labels were always stranded in columns 1 to 5, possibly far from the DO termination to which they referred. An example is shown below:

```
        DO 100 I = 1,L
          DO 200 J = 1,N
            X = 0.0
              DO 300 K = 1,M
                 (Various statements)
300                CONTINUE
                 (Various statements)
200              CONTINUE
               (Various statements)
100        CONTINUE
```

Obsolescent features of FORTRAN 77

In the DO statement

```
DO label VAR = EXP1, EXP2, EXP3
```

it was permissible for VAR, EXP1, EXP2 and EXP3 to be INTEGER, REAL or a mixture. In the last case, TYPE conversions were made before assigning VAR a value, as described in Chapter 3. Thus, assuming implicit typing of variables, one could write

```
DO 20 X = 1.2,4.5,1.4
```
or
```
DO 30 K = 1.2,4.5,1.4
```

but the resulting number of loop completions could be unexpected (the interested reader could experiment using different processors). In Chapter 2 we saw that a REAL X of 4.5, as stored in a computer, was rarely "equal" to 4.5 and ran the risk of rounding on conversion to an INTEGER. Therefore it is strongly advised that non-integer DO loop counting should be avoided.

As was mentioned above, it was possible in FORTRAN 77 to terminate one or more DO loops at an executable statement (rather than a CONTINUE). For example

```
      DO 100 I = 1,L
       DO 100 J = 1,N
        DO 100 K = 1,M
100       (Do something)
```

This can also have unexpected consequences and should be avoided.

A further variation of the DO construct available in extended versions of FORTRAN 77 took the form

```
       DO label WHILE (logical expression)
         .
         .
         .
         .
    label CONTINUE
```

Since this is entirely equivalent to the endless DO construct

```
DO
if (logical expression is not true) EXIT
  .
  .
  .
END DO
```

the 'DO WHILE' construct could be considered to be obsolete. For the precise forms of "logical expressions" in Fortran 90 see Chapter 6 which follows.

Exercises

1. Write a program which reads in n REAL numbers (which we will call $x_1, x_2, x_3, \ldots, x_n$) and calculates the following statistics:

   (a) Their sum, i.e. $\sum_{i=1}^{n} x_i$, which we will call SUM
   (b) The sum of their squares, i.e. $\sum_{i=1}^{n} x_i^2$, which we will call SUMSQ
   (c) Their arithmetic mean, i.e. SUM/n
   (d) Their standard deviation, i.e.

$$\sqrt{\frac{\text{SUMSQ} - \dfrac{\text{SUM}^2}{n}}{n-1}}$$

   (e) The largest and smallest numbers
   (f) The "range", i.e. the difference between the largest and smallest numbers

   Hint: Use the intrinsic procedures from Chapter 4.
   Test your program on the set of five positive numbers 5.67, 11.38, 9.27, 10.56, 15.23.

2. Write a program to compute

$$\int_a^b \frac{\sqrt{y}\sin y}{y + e^y}\,dy$$

   using the "trapezoidal" rule (i.e. by approximating the function by linearly varying small steps). Read in a, b and n, the number of subdivisions of the range a to b and output the value of the integral for the range 0 to $\pi/2$ using 10 and 100 subdivisions respectively.

3. The cosine of an angle x, in radians, is the result of the summation of the infinite series

$$1 - \frac{x^2}{2!} + \frac{x^4}{4!} - \frac{x^6}{6!} + \cdots$$

   Write a program to read in a number of angles, and for each of them to sum the first 50 terms of the series, printing out the intermediate sum after each 5 terms. At the end, compare your summations with the intrinsic procedure value COS(X). Test the program for angles of 30°, 570° and 1470° and comment on the results.

6
Conditions

Introduction

The programs we have seen so far have mainly consisted of statements executed in sequence, one after the other. By means of the DO loop, we have been able to execute some groups of statements repetitively, thus interfering with the natural sequential process. In general, we shall need more widespread power to intervene in the course of a program's execution and to do various things depending on the states of variables at that particular juncture. This involves the capacity to choose what to do next according to whether the answer to a question is true or false, and the language provides a special TYPE of variable which can only have one or other of these two values or states (like an on-off switch).

Variables and constants of TYPE LOGICAL

Logical or "Boolean" variables can be declared in Fortran 90 programs by declarations of the form

```
LOGICAL : : ON,OFF
```

Variables such as ON and OFF can only take the values .TRUE. or .FALSE. where the English words are bracketed by decimal points. Thus statements like

```
ON = .TRUE.
OFF = .FALSE.
```
can be written.

Shorthand versions of these rather cumbersome logical values could be achieved by a declaration such as

```
LOGICAL, PARAMETER : : T = .TRUE., F = .FALSE.
```

making T and F logical constants.

Logical operators

Logical variables can be operated on by a small set of operators the simplest of which, .NOT., negates or changes the logical value of the single variable it operates on. Therefore if logical variable ON has the value .TRUE. as above, .NOT. ON has the value .FALSE..

The remaining logical operators are "binary" in nature in that they are placed between two logical variables and used to compare the "states" of these variables. They are: .AND., .OR., .EQV. and .NEQV..

The order of precedence of the logical operators is .NOT., .AND., .OR., .EQV. and .NEQV.. Parentheses would have to be used to override the normal order of precedence. Left to right evaluation of expressions is carried out as it was in arithmetic expressions.

The difference between the "exclusive" .AND. and the "inclusive" .OR. can be appreciated with respect to the logical variables ON and OFF declared above. The result of

```
ON .AND. ON
```

is .TRUE. because both are .TRUE. but the results of

```
OFF .AND. OFF
```

and of

```
ON .AND. OFF
```

are .FALSE. because at least one expression in each case is .FALSE..

On the other hand the results of

```
ON .OR. ON
```

and of

```
ON .OR. OFF
```

are .TRUE. because at least one expression in each case is .TRUE., whereas only the result of

```
OFF .OR. OFF
```

is .FALSE. because both are .FALSE.. Complicated logical operations are to be avoided because it is very easy to become confused as to the resulting states of variables.

Logical variables cannot be tested to see for example if they are "equal". Two logical variables in the same state would be said to be "equivalent", hence the .EQV. and .NEQV. operators.

A complete "truth table" can be drawn up as below, summarising the results of the binary operations on the LOGICAL variables ON and OFF

ON	OFF	ON.AND.OFF	ON.OR.OFF	ON.EQV.OFF	ON.NEQV.OFF
.TRUE.	.TRUE.	.TRUE.	.TRUE.	.TRUE.	.FALSE.
.TRUE.	.FALSE.	.FALSE.	.TRUE.	.FALSE.	.TRUE.
.FALSE.	.TRUE.	.FALSE.	.TRUE.	.FALSE.	.TRUE.
.FALSE.	.FALSE.	.FALSE.	.FALSE.	.TRUE.	.FALSE.

Relational operators

In order to make choices, we have to have the ability to compare the results of numeric (and other) expressions and act accordingly. For this purpose, Fortran 90 provides a set of "relational operators" which are placed between expressions in much the same way as arithmetic operators are placed between variables in arithmetic expressions. The six relational operators and their interpretations are:

<	less than
< =	less than or equal to
= =	equal to
/ =	not equal to
>	greater than
> =	greater than or equal to

Logical expressions

Using the above relational operators we can create logical expressions whose values are either .TRUE. or .FALSE.. A simple form of logical expression has the form:

   arithmetic expression   relational operator   arithmetic expression

Using this form we can compare expressions to see if their resulting values are equal or not. For example the logical expression to compare the REAL variables A and B

   A < B

has the value .TRUE. if A is less than B, otherwise it has the value .FALSE.. Care is obviously necessary when testing for equality (see Chapter 2) in that

   A = = B

will rarely be .TRUE. for REALs A and B whereas

   I = = J

is a perfectly reasonable test for equality of INTEGERs I and J.

A more general logical expression has the form:

logical expression   logical operator   logical expression

and using this form we can test the result of

```
A < B .AND. A < C
```

which will be .TRUE. if A is less than both B and C, otherwise it will be .FALSE..

Logical ("IF") statements

Now that we know how to make choices on condition that a certain state is .TRUE. or .FALSE. we can select which sections of code are to be obeyed depending on the circumstances. This is sometimes called "branching".

The simplest variation of the logical statement has the form

IF (logical expression) Fortran 90 statement

'Logical expression' can only have one of the two states .TRUE. or .FALSE. and the consequence of the above conditional statement is that the Fortran 90 statement would be obeyed if the logical expression gives the result .TRUE. but would not be obeyed if the logical expression gives the result .FALSE..

Thus

```
IF (A < B) B = A
```

results in B being assigned the value A only if it is numerically greater than A. Note that the range of numerical values in Fortran 90 is from − large number to + large number so that $5.27 > -263.9$ on this scale.

In this very simplest IF statement, the Fortran 90 statement which is conditionally executed is a simple statement and cannot be a "block" statement (see below), another simple IF statement, or END.

It is important to get used to assignment statements of the form

LOGICAL variable = LOGICAL expression

For example in some repetitive arithmetic process, "convergence" may be achieved when a measure of error is less than a small quantity, say 10^{-6}. Having declared a REAL variable ERROR and a LOGICAL variable CONVERGED we can say

```
CONVERGED = ABS (ERROR) < 1.E-6
```

and then

```
IF (CONVERGED) do something.
```

The block IF statement

A more usual branching requirement than the above is to be able to execute several statements (a "block" of statements) depending on some condition, or to execute different blocks depending on the circumstances.

The simplest conditional statement involving a block takes the form:

```
NAME:   IF (logical expression) THEN
            Statement 1
               ⋮
            Statement n
        ENDIF NAME
```

For example:

```
CHANGE:   IF (A < B) THEN
              STORE = A
              A = B
              B = STORE
          ENDIF CHANGE
```

which reverses A and B if A is less than B.

As with DO loops, the name preceding an IF block is optional but if it appears, the same name must be used after ENDIF.

The above is really a special case of the fuller block IF structure which takes the form:

```
NAME:   IF (logical expression) THEN
            Statement 1
               ⋮           First block of
                           statements
            Statement n
        ELSE
            Statement a
               ⋮           Second block of
                           statements
            Statement z
        ENDIF NAME
```

This causes all of the first block of statements to be executed if the logical expression has the value .TRUE. but if it has the value .FALSE. then the second block of statements is executed.

For example

```
CHANGE_SIGN:   IF (A/ = B) THEN
                   A = −A
               ELSE
                   B = −B
               ENDIF CHANGE_SIGN
```

will result in a change of the sign of A if A is not equal to B otherwise the sign of B will be changed.

As has repeatedly been emphasised, this would be dangerous if A and B are REAL variables. A better course of action would be to declare a REAL constant, say TOL, which represents how close A has to be to B for them to be considered "equal". Then, using the intrinsic FUNCTION ABS (Chapter 4) one could write

```
REAL, PARAMETER : : TOL = 1 . E - 6
      _
      _
      _
CHANGE_SIGN : IF (ABS (A — B) > = TOL)
            etc
```

The final and most complete form of block IF construction allows nesting of the IF ... THEN ... ELSE ... form to any number of levels as follows:

```
NAME:   IF (logical expression 1) THEN
           block 1
        ELSE IF (logical expression 2) THEN
           block 2
        ELSE IF (logical expression 3) THEN
           block 3
        ELSE
           block 4
        ENDIF NAME
```

Since the blocks can themselves be nested block IF constructions, they can also be named and an indented layout is essential to keep track of the program structure. For example

```
OUTER:   IF (I < J) THEN
            INNER:   IF (K = = L) THEN
                        P = Q
                        R = S
                        T = U
                     ENDIF INNER
         ELSE IF (M > N) THEN
            OTHER:   IF (K/ = J) THEN
                        P = — Q
                        R = — S
                        T = — U
                     ENDIF OTHER
                     ELSE
            Q = (P + R + T) * .5
         ENDIF OUTER
```

The structure of nested block IF statements

Whereas nested DO loops had a rather simple structure (for every DO there must be an END DO) the same is not true of nested IF statements and it is possible for these to become quite confusing even in simple programs. In order to determine which block IF, END IF, ELSE and ELSE IF statements "correspond" with one another, the concept of "IF level" can be helpful.

The "IF level" of a statement, say S, is defined as N1–N2 where

• N1 is the number of block IF statements from the beginning of the program unit up to, and including, statement S and

- N2 is the number of END IF statements in the program up to, but not including, statement S.

An END IF, ELSE or ELSE IF statement is said to "correspond" with a block IF statement if it is the next one having the same IF level as that block IF.

For example, the following outline program shows some IF levels and correspondences

Statement number	Statement	IF-level
1	IF ( ) THEN	1
2	...	1
3	END IF	1
4	...	0
5	IF ( ) THEN	1
6	...	1
7	IF ( ) THEN	2
8	...	2
9	ELSE	2
10	...	2
11	IF ( ) THEN	3
12	...	3
13	END IF	3
14	...	2
15	END IF	2
16	...	1
17	ELSE IF ( ) THEN	1
18	...	1
19	ELSE IF ( ) THEN	1
20	...	1
21	ELSE	1
22	...	1
23	END IF	1

It can be seen that statements 9 and 15 correspond to statement 7, the IF block beginning at statement 5 consists of all statements up to statement 16 and so on. If the LOGICAL expression in statement 5 is .TRUE. statements 6 to 16 inclusive will be executed followed by statement 23.

CASE conditions

It is often convenient to be able to select branches in a program according to a very simple criterion, for example that an INTEGER is 1,2,3,4, or in a certain range, or whatever. (This particular variant was called the "computed GO TO" in FORTRAN 77 but in that form it now joins the list of obsolescent features—see below).

The form of the CASE expression is:

```
NAME:   SELECT CASE (expression)
            CASE (Selector)
            Block
            CASE (Selector)
            Block
            etc
            CASE DEFAULT
            Block
            END SELECT NAME
```

For example

```
GAUSS_POINTS: SELECT CASE (NGP)
  CASE (1)
    ABSCISSA = .0
    WEIGHT = 2.
  CASE (2)
    ABSCISSA = 1./SQRT(3.)
    WEIGHT = 1.
  CASE (3)
    ABSCISSA = .2 + SQRT(15.)
    WEIGHT = 5/9.
  END SELECT GAUSS_POINTS
```

In the above example the discriminating quantity is the INTEGER NGP but in general LOGICAL, CHARACTER (see Chapter 7) or INTEGER expressions can be used but not REAL. Ranges of selectors can be employed (1:5 means all the INTEGERs from 1 to 5 and :-1 means all the negative INTEGERs). The selectors cannot be variables or expressions.

For example a year represented as the value NYEAR could be located by decade according to the following selection where DECADE is a CHARACTER variable (see Chapter 7):

```
DECADES : SELECT CASE (NYEAR)
          CASE (:1899)
            DECADE = 'PAST'
          CASE (1900:1909)
            DECADE = 'HUNDREDS'
          CASE (1910:1919)
            DECADE = 'TEENS'
          CASE (1920:1929)
            DECADE = 'TWENTIES'
          CASE (1930:1939)
            DECADE = 'THIRTIES'
                 :
          CASE (1990:1999)
            DECADE = 'NINETIES'
          CASE DEFAULT
            DECADE = 'FUTURE'
          END SELECT DECADES
```

Leaving DO loops via EXIT or CYCLE

We are now in a position to give the full form of the 'endless' DO constructions we saw in Chapter 5. The following program calculates the average of a series of input positive REAL numbers, terminating when a negative number is input

```
PROGRAM MEAN
   ! this program finds the mean of
   ! a set of input data, one number per record
     IMPLICIT NONE
     REAL : : QUANTITY, SUM
     INTEGER : : NO_OF_VALUES_INPUT
       SUM = 0.0
       NO_OF_VALUES_INPUT = 0
         DO
           READ*, QUANTITY
           IF (QUANTITY < = .0) EXIT
           SUM = SUM + QUANTITY
           NO_OF_VALUES_INPUT = &
           NO_OF_VALUES_INPUT + 1
         END DO
         PRINT*, "The Mean of the Numbers Input is", &
         SUM/NO_OF_VALUES_INPUT
   END PROGRAM MEAN
```

The following program calculates the number of positive odd integers, from a series of integers input one per record, until input is terminated by a negative integer.

```
PROGRAM EVEN_OR_ODD
   ! this program finds the number of odd
   ! integers in a mixed list of even and odd
   ! values.
  IMPLICIT NONE
  INTEGER : : NO_OF_ODDS, NUMBER_INPUT
    NO_OF_ODDS = 0
    DO
      READ*, NUMBER_INPUT
      IF (NUMBER_INPUT < 0) THEN
        EXIT
      ELSE IF (MODULO (NUMBER_INPUT, 2) = = 0) THEN
        CYCLE
      ELSE
        NO_OF_ODDS = NO_OF_ODDS + 1
      END IF
    END DO
    PRINT*, "The number of odd numbers found is", NO_OF_ODDS
  END PROGRAM EVEN_OR_ODD
```

Structure chart representations

The basic form of choice is represented by the branching box structure below:

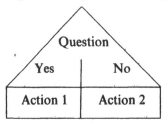

which can easily be extended to the IF . . . THEN . . . ELSE IF . . . THEN . . . ELSE . . . ENDIF form as:

Question					
Answer 1	Answer 2	Answer 3	Answer 4	Answer 5	Answer 6
Action 1	Action 2	Action 3	Action 4	Action 5	Action 6

As an example, the program listed below carries out various operations, depending on the value of an INTEGER N which is read in. If N is less than zero, an error is reported. If N is greater than 10 the INTEGER part of N/10 is printed. If N lies between 1 and 9 the real quantity N/10 is computed and printed unchanged or multiplied by 10 according to its size:

```
PROGRAM STRUCTURE
! illustration of structure charts
 IMPLICIT NONE
 INTEGER : : N
 REAL : : QUANTITY
   READ*,N
    IF (N < 0) THEN
      PRINT*,'ERROR'
    ELSE IF (N > 10) THEN
       PRINT*,N/10
    ELSE
       QUANTITY = N/10
       IF (QUANTITY < .5) THEN
         PRINT*,QUANTITY
       ELSE
         PRINT*,QUANTITY*10
       ENDIF
    ENDIF
END PROGRAM STRUCTURE
```

The corresponding program structure chart would be:

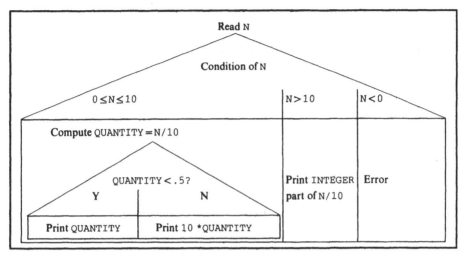

It can be seen that the convention is to enclose the whole program by a rectangular box.

Differences from FORTRAN 77

The more concise relational operators < etc have replaced their more cumbersome FORTRAN 77 equivalents as follows:

```
<       replaces .LT.
< =     replaces .LE.
= =     replaces .EQ.
/ =     replaces .NE.
>       replaces .GT.
> =     replaces .GE.
```

IF statements are essentially unchanged but can be labelled to improve program documentation.

Obsolescent features of FORTRAN 77

The SELECT CASE facility replaces the "computed GO TO" in FORTRAN 77. The GAUSS_POINTS example given above used to take the form:

```
        GO TO (1,2,3), NGP
  1     SAMP1 = .0
        SAMP2 = 2.
        GO TO 100
  2     SAMP1 = 1./SQRT (3.)
        SAMP2 = 1.
        GO TO 100
  3     SAMP1 = .2*SQRT (15.)
        SAMP2 = 5./9.
100     CONTINUE
```

which is now thought to be obsolescent. Since all of FORTRAN 77 is contained within Fortran 90, the above "unconditional GO TO" statement GO TO 100 is still legal and just means "go to statement labelled 100". However, as was the case with DO loops, the absence of labels in the new SELECT CASE construct means that labels are scarcely necessary at all in Fortran 90 programs and the language can be taught in the first instance without mentioning "GO TO" at all.

As was illustrated in the previous chapter, the "DO WHILE" construct can be replaced by an "endless DO" as in the following example. The sequence

```
DO 100, WHILE ( (B**B-4.*A*C).GE.0.)
    ⋮
100 CONTINUE
```

is better expressed as:

```
REAL_ROOTS:  DO
                ⋮
             IF ( (B*B-4.*A*C) < = 0.) EXIT
                ⋮
             END DO REAL_ROOTS
```

A regrettably common form of conditional statement in FORTRAN 77 programs is the "arithmetic IF" which contains three labels as follows:

```
IF (arithmetic expression) label 1, label 2, label 3.
```

This will result in program execution jumping to the statement labelled "label 1" if the arithmetic expression is less than zero, to "label 2" if the arithmetic expression is equal to zero, and to "label 3" if the arithmetic expression is greater than zero. Given that we seek as far as possible to avoid labels in Fortran 90, the "arithmetic IF" must be considered obsolescent.

Less frequently used, but even more regrettable constructions in FORTRAN 77 involve the "assigned GO TO" and ASSIGN statements. If K is an INTEGER variable, it was possible to write

```
ASSIGN label TO K
        etc
```

followed by the "assigned GO TO" in the form

```
GO TO K (label 1, label 2, label 3,... label N)
```

as long as one of label 1 to label N was equal to label. Control would then be transferred to the statement labelled K.

Alternatively, after an ASSIGN statement such as the one above, K could be the label of a FORMAT statement (see Chapter 7).

All of these uses of ASSIGN and its associated "assigned GO TO" are considered to be obsolescent and should not be used in Fortran 90 programs.

A final, possibly little-used feature of FORTRAN 77 which is obsolescent, is that it was possible to jump to an END IF statement from outside the corresponding "IF" block. Now control should be passed to the statement immediately following the IF block.

Exercises

1. Write a program to read in three numbers a, b and c, assumed to be the coefficients of the quadratic

$$ax^2 + bx + c = 0$$

and calculate and print the roots. Account for the three possibilities

(a) two different real roots

(b) one (double) root

(c) a pair of complex conjugate roots.

2. Consider a number spiral drawn on the coordinate axes, as shown below (Fig. 1). This is produced by placing 0 at the origin, 1 immediately to its right at (1,0), 2 immediately above 1 at (1,1), 3 immediately to the left of 2 at (0,1), and so on.

Write a program which will read in a number and print out its coordinates on the number spiral.

Hints: The following observations might be useful. Consider the spiral as made up of concentric rings numbered, from the centre, 0,1,2...r. In Fig. 2, ring 2 is cross-hatched. Within ring r are contained numbers between

$$(2r - 1)^2 \text{ and } (2r + 1)^2 - 1.$$

Thus, given a number, its ring can be found.

36	35	34	33	32	31	30	
37	16	15	14	13	12	29	
38	17	4	3	2	11	28	
39	18	5	0	1	10	27	
40	19	6	7	8	9	26	51
41	20	21	22	23	24	25	50
42	43	44	45	46	47	48	49

Fig. 1 A number spiral

Which of the four arms of the ring the number lies on can be found by noting that, for ring r, the corners are occupied by the numbers $2r(2r - 1)$, $2r(2r)$, $2r(2r + 1)$, $2r(2r + 2)$.

Finally, the precise position of the number can be found.

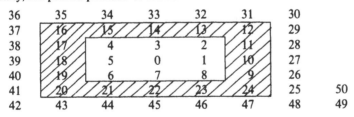

Fig. 2 Ring 2 (cross-hatched) of the number spiral

By observing the pattern of the locations of squared numbers, predict where 625 would be found and check using your program.

3. In a survey conducted by an opinion poll, voters were asked two questions

(a) How do you intend to vote in the forthcoming election?

(b) How did you vote last time?

In both cases, the possible answers were "Conservative", "Labour", "Liberal- Democrat" and "Others" in the United Kingdom and "Republican", "Democrat", "Maver-

ick" and "Others" in the United States. The answers were written down on a question-
naire (one per voter).

The responses are to be analysed by computer. The information to be extracted is
the percentage who intend to vote in each of the four categories, and, in the case of
intending Liberal-Democrats (Mavericks) *only* who have changed their allegiance
since last time, the percentage of them who have changed from Conservative
(Republican) and from Labour (Democrat).

One hundred and forty four people were interviewed. The answers were stored as
two numbers, using the code 1 = Conservative (Republican), 2 = Labour (Democrat),
3 = Liberal-Democrat (Maverick), any other number = Others. Twelve responses
could be stored per record (line).

Write a program to do the analysis, given the data below:

```
1 1  2 2  1 5  3 1  3 2  3 3  1 1  2 2  4 1  1 2  2 1  5 1
4 1  4 2  1 1  2 2  3 3  1 1  2 2  1 5  1 1  2 1  2 2  3 3
5 2  3 2  3 1  4 1  2 2  1 1  3 2  5 1  1 2  1 1  1 1  2 2
7 1  2 1  2 2  1 2  1 1  1 1  2 1  1 2  4 1  3 2  3 1  1 1
1 1  1 1  2 2  2 2  2 2  3 1  1 2  1 1  2 2  3 3  1 2  1 2
3 1  1 1  1 1  1 2  1 2  3 1  3 3  1 3  1 1  1 1  1 1  1 1
2 2  2 2  2 2  3 1  1 1  1 1  2 2  2 1  1 1  1 5  1 1  1 1
3 2  3 3  4 1  1 2  1 1  2 1  1 1  1 1  2 2  2 2  2 2  1 1
6 2  1 3  3 1  1 2  1 1  2 2  2 2  2 2  3 1  3 2  1 1  2 2
4 1  1 4  3 2  3 3  1 2  2 2  2 2  2 2  1 2  1 2  2 1  2 1
2 2  3 3  3 3  2 2  1 1  2 1  2 2  2 2  1 4  1 4  4 1  4 2
1 3  1 1  2 2  1 1  2 2  1 1  2 1  1 1  1 1  1 1  4 3  3 4
```

4. Stability of a Concrete Dam

The figure below illustrates force equilibrium for a rigid block held against a rigid base
by its own weight W while being pushed to the right by an external force F.

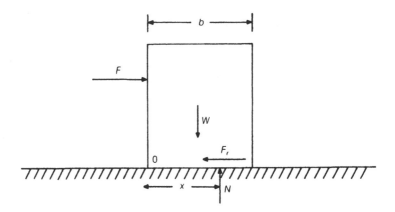

If the block is homogenous, W acts through the centroid. A frictional force F_r is
generated as the block tries to slide over the base. The maximum value that F_r can
attain is the "sliding resistance".

$$F_{rmax} = \mu N \qquad \text{where } \mu = \text{coefficient of static friction}$$

Stability against sliding

Any design must ensure that $F_r < F_{rmax}$ to prevent the block sliding to the right across the rigid foundation.

Stability against toppling

The position (x) of the resultant normal force N must be found by taking moments. Should the point of action of N move outside the block $(x > b)$, toppling failure will result.

These principles can be used to design the dam shown below, provided it stands on a rigid base

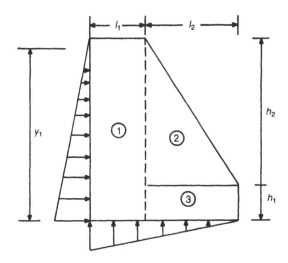

The dam is made of concrete, which weighs 24 kN/m³, and holds back the hydrostatic (triangular) pressure of water to a height y_1. Because water tends to seep between the dam and its foundation a triangular "uplift" pressure is shown acting under the dam. As a conservative estimate, the uplift pressure is assumed to vary linearly from the hydrostatic pressure at the "upstream" (left) end of the base to zero at the "downstream" (right) end of the base. The influence of the "uplift" pressure is to reduce the effective weight W of the dam.

(a) Develop the logic of a computer program to analyse the stability of any dam having the above general geometry (you may assume constant conditions along the "length" of the dam, normal to the drawing, so that a typical 1m length can be analysed). It will be convenient to divide the cross-section of the dam into three areas as shown.

(b) Write a program to analyse the dam's stability against sliding and toppling.

(c) As an additional constraint, it is usual to arrange for the normal force N to act within the "middle third" of the width of the base, i.e. $b/3 \geqslant x \geqslant 2b/3$. Add this condition to your program.

(d) Check whether the following dam is stable:

$$l_1 = 12\,\text{m}$$
$$l_2 = 18\,\text{m}$$
$$h_1 = 3\,\text{m}$$
$$h_2 = 27\,\text{m}$$
$$y_1 = 24\,\text{m}$$
$$\mu = 0.4$$

5. Modify the program written for Exercise 3 of Chapter 5 so that the summations are stopped when the difference between successive sums is "small enough".

7

More on Input/Output: CHARACTERS and FORMAT Specifications

Introduction

So far we have only seen how to input REAL and INTEGER variables from the keyboard and how to print them out on the screen using simple "list-directed" input/output statements like

```
READ*, A,B,C
PRINT*, I,J,K
```

We have also seen that PRINT can be used to display text or "characters".

In due course we shall want to read from and write to various sources called FILEs, but before describing facilities to do this, see Chapter 9, we first describe how to manipulate input and output of text or "characters", rather than numerical data, and how to arrange numerical data, particularly on output, in a neat presentation using FORMATs.

CHARACTER variables

Text can be represented in Fortran 90 by most of the characters on the typical keyboard—the alphanumerics together with $+$, $-$, $=$ and so on. In order that text should be recognised as such, it must be enclosed in single or double quotes, for example

'This is an Example of Text'.

In order for text to be manipulated readily, a special TYPE of variable, the CHARACTER variable, can have text assigned to it. We can declare

```
CHARACTER (LEN = 20) : : TEXT
```

where LEN is the maximum number of characters in the string, and make assignments such as

```
TEXT = 'Dollars and Cents'
```

as long as the CHARACTER variable is enclosed in quotes. Note that blanks are treated as such—there is no need for the underscore symbol which was necessary in ordinary variable names to simulate a space. A string of more than 20 characters will be truncated.

In order to be able to include quotes in a CHARACTER variable, the double or single quotes can be used as the delimiters of the variable, for example

```
"THIS DOESN'T MATTER"
```

is acceptable, as is

```
'NEVER SAY "NEVER" '
```

It may however be a good idea for programmers to standardise their usage to either single or double quotes when a CHARACTER variable is intended.

The above declaration of the CHARACTER variable TEXT assumes that the number of characters in the variable will never exceed 20. The declaration can be abbreviated to

```
CHARACTER (20) :: TEXT
```

The declaration of CHARACTER constants follows the same rule as for the REAL, INTEGER, COMPLEX and LOGICAL TYPEs we have seen so far, for example

```
CHARACTER (1), PARAMETER :: QUERY = '?'
```

The default length of a CHARACTER variable or constant is 1, as in the example

```
CHARACTER :: DIGIT
```

When the number of characters in a CHARACTER variable is not known until run time (in the context of a subprogram—see Chapter 10), we may use declarations of the type

```
CHARACTER (*) :: FIRST_NAME, SURNAME
```

List-directed input and output of character variables can then be used, just as it was for numerical data. With the definition of TEXT given above

```
PRINT*, TEXT
```

or

```
PRINT*, 'Dollars and Cents'
```

result in the expected display on the screen.

We can now see in principle how to output results in a more meaningful way. Character and numerical variables can be included in the output, separated by commas, so that one could say:

```
PRINT*, 'Pay for Week ending', I, 'May is', D, 'Dollars'
```

where INTEGER variable I contains the day of the month and REAL variable D the amount to be paid.

There is a special character string concatenation operator, //, which may be used to combine strings into longer ones, for example

```
'Ms.' // FIRST_NAME // SURNAME
```

assuming FIRST_NAME and SURNAME are CHARACTER variables.

There is also a series of intrinsic CHARACTER FUNCTIONs (see Appendix 3). For example the intrinsic procedure LEN returns the length of its CHARACTER argument. Thus LEN ("This") is 4. Since a CHARACTER variable often does not fill the space allotted to it, the FUNCTION called TRIM removes all trailing blanks. In the above examples PRINT*, TRIM (TEXT) would output just the variable required.

Characters have a hierarchy or "collating sequence" established by ISO standards. For example the ASCII hierarchy is

blank ! " # $ % & ' ( ) * + , − . /
0 1 2 3 4 5 6 7 8 9 : ; < = > ? @
A B C D E F G H I J K L M N O P Q R S T U V W X Y Z [ ] ˆ _ '
a b c d e f g h i j k l m n o p q r s t u v w x y z { | } ˜

A character is "less than" another character if it precedes it in the sequence, and this can be tested using the logical operators given in Chapter 6.

The values of CHARACTERs in given positions of the collating sequence can be found by the intrinsic FUNCTIONs ACHAR(I) for the ASCII sequence and CHAR(I) for the user's particular processor where I is the position. On a given system, the sequence need not start at 1, for example the blank may be positioned at I = 32. This can be found by the position intrinsic FUNCTIONs IACHAR(C) and ICHAR(C) for ASCII and processor respectively where C is a CHARACTER variable. Thus IACHAR(' ') may return 32 but thereafter the characters will be in sequence.

Substrings

Another useful intrinsic procedure is INDEX which returns the position of the first occurrence of one CHARACTER string as a "substring" of another. For example

```
INDEX ("Shakespeare", "spear")
```

returns 6. If a substring occurs more than once, it is its first occurrence which is returned, for example

```
INDEX ("concatenation", "on")
```

returns 2. If no occurrence of the substring is found, INDEX returns 0.

A contiguous sequence of CHARACTERs in a string can be referred to in a convenient way by two INTEGER expressions giving the positions of the first and last

CHARACTERs in the substring. The integer expressions are enclosed in parentheses and separated by a colon. Thus "Shakespeare" (7:10) is "pear".

FORMAT specifications

In list-directed input/output the asterisk symbol, *, really stands for "free" format, that is, not under the control of the programmer. Indeed it is shorthand for FMT = * where FMT is normally a CHARACTER variable or expression.

We saw that when a REAL variable was input as 1.9 the same variable could well be output as 1.89999962.

If we want to control the format of the output, so that 1.89999962 appears on the screen as 1.9, or indeed in one of the other valid forms shown in Chapter 3, such as

1.900
1.9E + 00
.19E + 01

we must convert the internally represented number to its displayed form by means of a "format specification".

Such specifications define a "field" within which the characters representing the number will be displayed. As a simple example, suppose the INTEGER number 56 is to be displayed in a "field" of 8 characters. The "width" of the field is said to be 8. Using **b** to represent a blank character, there are clearly many ways in which the number could appear:

bb56bbbb
bbbbbb56

and so on. The FORMAT specification of this field in Fortran 90 is I8 (an INTEGER field, 8 characters wide) and the convention is that the number would be "right-justified" in the field as in the second of the two examples given above.

There are numerous "FORMAT descriptors" like the above "I" example but we shall limit ourselves to a discussion of it and only three more, namely the "F" and "E" descriptors for displaying REAL numbers, and the "A" descriptor for displaying CHARACTERs.

I FORMAT

The example given above is a particular case of the general Iw form for representing INTEGERs where w is a non-zero, unsigned integer defining the width of the field. As stated above, the number will be right-justified in the field and may be signed so that −57 printed out in I5 specification would appear as **bb** − 57.

There is an alternative output form Iw.m where w is as always the field width but at least m digits are to be printed. Under I5.3 control, −57 would now appear as **b** − 057.

A danger of formatting is immediately apparent. What if we try to represent an INTEGER which is too big to fit the field? On output, the field will be filled with asterisks. On input, the right-justification rule can mean that if 100000 is input to an I5 field, the number represented internally will be zero. If left justified it would be 10000.

F FORMAT

The general form is Fw.d, where w and d are unsigned integers which define the field width and number of digits after the decimal point respectively. The decimal point counts as a position in the field. Thus 8.27 in F7.3 specification would appear as **bb** 8.270. Clearly there are potential dangers again—on input the decimal point may be mis-positioned (in which case the value of d is ignored) and on output the variable may be too large to fit the field. Truncation and rounding then take place which are extremely dangerous. For example the variable with value 56.265 is output as 56.26 if the format specification is F5.2.

E FORMAT

It will be clear from the preceding section that unless the range of REAL variables to be output is very rigidly pre-defined, the danger of "fixed point" F specification output leading to erroneous results is unacceptably high. Most engineering programs therefore use "floating point" or E specification for REALs which in the most common form is written Ew.d. As usual, w is the width of the field and d the number of digits following the decimal point. As an example, the variable 10.0, when output by an E12.4 specification, would appear as **bb**.10000E+02. Since the "exponent" part, E+02 in the above example, allows only two digits after the sign, this form is clearly limited to exponents less than or equal to 99.

Scaling in engineering calculations usually means that very large exponents can be avoided. If this is impossible, there is another form of E specification Ew.dEe where e is an unsigned non-zero integer determining the number of digits to appear in the exponent field. Thus, with the specification E12.4E4 we could represent 1.5×10^{1000} as **b**.1500E+1001.

A FORMAT

The form is A or Aw where as usual w is the field width. If it is omitted, the A specification is adjusted to fit the width of the variable to be input or output. For example the CHARACTER variable 'LENGTH' could be output under A or A6 as LENGTH or under A8 as **bb**LENGTH. If the field is too small, the variable is left justified, so that 'LENGTH' under A3 would appear as LEN.

Repeats in specifications

If more than one quantity is to be input or output with exactly the same FORMAT, repeat counts are allowed. For example 5I6 FORMAT signifies five consecutive

INTEGER fields six characters wide. Groups of fields can be repeated, by enclosing them in parentheses, thus 5(I6,F6.3) signifies five consecutive fields 12 characters wide, each made up of an INTEGER and a fixed point REAL. Further nesting, e.g. 2(3(F5.2, I2)) has the obvious implication.

Use of FORMAT specifications

Now that we know some simple FORMAT specifications, we can use them to replace the list-directed or "free" FORMAT statements we have used so far. Assuming IMPLICIT typing we could replace

```
PRINT*,I,A,B
```

by

```
PRINT'(I5,F6.2,E12.4)', I,A,B
```

which would give us direct control over the appearance of the variables I, A and B on the screen. Due to the perils of formatting described above, it is suggested that formatted input be avoided by beginners. In the above example, the CHARACTER string '(I5,F6.2, E12.4)' is really shorthand for FMT='(I5, F6.2, E12.4)' but we return to a more detailed discussion of input and output in Chapter 9.

The use of A FORMAT is necessary when interspersing text with numerical values of variables (which we did quite naturally using "free" FORMAT). For example

```
PRINT*, 'The value of x is', X
```

becomes

```
PRINT '(A,F12.4)', 'The value of x is', X
```

In list-directed output, a blank is output at the beginning of each record so that it starts on a new line. If, however, an output list is too long for its FORMAT, it will be output on new lines in that FORMAT until the output list has been exhausted.

Useful edit descriptors

There are several facilities for controlling the appearance of output but the most commonly used are those allowing the insertions of blank spaces, the so-called "X" descriptor, and those for starting a new record, the so-called "/" descriptor.

Thus, output could be arranged with blanks and new lines by a statement such as

```
PRINT'(2X,I5,F6.2,3/,E12.4)', I,A,B
```

"Tabbing" to the right using nX can also be achieved (better?) by TRn and it is possible to tab to the left using TLn or to position n using Tn.

Differences from FORTRAN 77

These are not significant in the above material. Mention should be made here of the FORMAT statement, but since this is a labelled statement, its use is discouraged in Fortran 90. It takes the form

    label FORMAT (format specification)

for example

    100 FORMAT (I5,F6.2,E12.4)

The corresponding PRINT statement would then take the form

    PRINT 100,I,A,B

Exercises

1. Write a program which will read in a pair of integers, say P and Q, and print out their highest common factor (HCF) and lowest common multiple (LCM). The HCF is often calculated using "Euclid's algorithm" the steps of which are as follows.

    Divide P by Q and take the remainder R. If R is zero, Q is the HCF, otherwise repeat the process with Q and R in place of P and Q, continuing until a zero remainder is found.

    For example

    $380/266 = 1$ remainder 114
    $266/114 = 2$ remainder 38
    $114/38 = 3$ remainder 0

    Try your program on the data pairs

    380 266
    266 380
     24 1

2. A Pythagorean triangle is a right-angled triangle whose sides are each an integer multiple of some fixed unit. Well known examples are triangles whose sides are in the ratios 3:4:5 and 5:12:13.

    Write a program to find all the Pythagorean triangles up to some given limit (say 100 units) of hypotenuse, by the following method.

    Let A be the hypotenuse and B and C the two other sides so that a successful solution has $A^2 = B^2 + C^2$. Since 5:4:3 and 5:3:4 are essentially the same solution let us take $B < C$. For each integer value of A, there is a trivial solution with $B = 0$ and $C = A$. We look for the non-trivial solution by increasing B by 1 or decreasing C by 1 until $B \geqslant C$. In general this does not give another right-angled triangle and $B^2 + C^2 = A^2 + \delta$. Only $\delta = 0$ gives a solution. When $\delta < 0$, B should be increased next time and when $\delta > 0$, C should be decreased. Whenever B is increased,

$$A^2 + \delta' = (B+1) + C^2$$

so

$$\delta' - \delta = (B+1)^2 - B^2 = 2B + 1$$

Whenever C is decreased

$$A^2 + \delta' = B^2 + (C-1)^2$$

so

$$\delta' - \delta = -(2C - 1).$$

By this means, changes in δ can be tracked.

Hint: Comparatively expensive operations such as taking square roots are not necessary.

3. A mortgage is an arrangement under which a person borrows a sum of money, say £A, from another person against a security such as a house. The loan is to be repaid over a fixed period, say n years, at the current rate of interest, say r per cent per annum. The annual repayment, say £P, is given by the formula

$$P = \frac{Ar\left(1 + \dfrac{r}{100}\right)^n}{100\left(\left(1 + \dfrac{r}{100}\right)^n - 1\right)}$$

Write a program which will print monthly mortgage repayment tables for some repayment period. The program should first read in the repayment period. It should then read in a range of annual interest rates, and a range of amounts borrowed, each in the form of a lower limit and an upper limit together with an increment. For each interest rate in the specified range, the program should print a monthly repayment table showing the monthly repayments corresponding to the specified range of amounts borrowed. For example, each table may look like this:

```
*****************************************************************
     MORTGAGE REPAYMENT TABLE          PAGE dd

    (REPAYMENT PERIOD = dd YEARS)
*****************************************************************
     ANNUAL INTEREST RATE = dd.dd%
*****************************************************************
   AMOUNT BORROWED                   MONTHLY REPAYMENT

     ddddd.dd                          ddddd.dd
     ddddd.dd                          ddddd.dd
     ddddd.dd                          ddddd.dd
        :                                 :
     ddddd.dd                          ddddd.dd
*****************************************************************
```

where d is a decimal digit. The data consist of three lines with the following layout:

Line no.	Contents
1	repayment period
2	lower limit, upper limit, and increment for interest range
3	lower limit, upper limit, and increment for loan range

4. Find all three-digit numbers which are equal to the sum of the cubes of their digits.

5. The "Mandelbrot" set is a catalogue of fractal shapes constructable from quadratic functions in the complex plane of the form $Z^2 + C$ where Z and C are complex. Write a program to square a complex number, add C, square the result and add C and so on, plotting the resulting set "graphically" using CHARACTER variables.

8

Arrays

East and west and south and north to summon his array
Macauley

Introduction

In previous chapters we have often referred to variables describing similar objects of the same TYPE as X1, X2, X3 etc and it is reasonable to expect scientific languages like FORTRAN to have the capacity to refer to all such objects by a shorthand notation comparable to the subscripted x_i form commonly used in mathematics. In this notation, x_i describes all the variables $x_{-\infty}, \ldots x_{-2}, x_{-1}, x_0, x_1, x_2 \ldots x_\infty$ collectively. Since there are no subscripts in the FORTRAN character set (Appendix 1), parentheses are used instead and the equivalent of x_i is written X(I). A "subscripted" variable such as X(56) is a (REAL) "element" of a "compound object" in Fortran 90 called an "array".

Array declarations

Whereas for simple variables of TYPEs REAL and INTEGER we had the option of whether to declare them or not, declaration of all array variables in a Fortran 90 program is obligatory. The declarations given below reserve space for a "one-dimensional" array of REALs called VECTOR and a "two-dimensional" array of INTEGERs called ARRAY respectively:

```
REAL,DIMENSION (10) : : VECTOR
INTEGER,DIMENSION (5,20) : : ARRAY
```

Alternative, and sometimes neater, equivalents are

```
REAL : : VECTOR(10)
INTEGER : : ARRAY(5,20)
```

The first of these declarations creates 10 REAL variables VECTOR(1), VECTOR(2), VECTOR(3) ...VECTOR(10) which are stored contiguously. The second creates 100 INTEGER variables ARRAY(1,1), ARRAY(1,2), ARRAY(2,1),...ARRAY(5,20) and there are two obvious strategies for storing them. In the first strategy ARRAY(1,2) would follow ARRAY(1,1) and in the second ARRAY(2,1) would follow ARRAY(1,1). The "rectangular" array ARRAY can be represented pictorially as follows:

ARRAY (1,1)	ARRAY (1,2)		ARRAY (1,20)
ARRAY (2,1)		Storage by columns	
↓	↓		
ARRAY (5,1)			ARRAY (5,20)

The convention is that the first subscript of a two-dimensional array is assumed to represent a row in the above matrix while the second subscript represents a column. In Fortran 90 implementations, storage is usually by columns, that is, ARRAY(2,1) follows ARRAY(1,1) and ARRAY(1,2) follows ARRAY(5,1) in store. However the Fortran 90 standard does not insist on this, and since the efficiency of many large engineering programs is directly affected, users are advised to check how their particular implementation works.

The numbers 10, 5 and 20 in the above declarations are examples of "upper bounds" on subscripts of array elements. We can see that "lower bounds" of 1 have been assumed but this is not essential. The ten numbers in array VECTOR above could be stored as VECTOR(-9:0) or in any other (usually) contiguous sequence. In general ARRAY(EAST:WEST, SOUTH:NORTH) is valid. Arrays of more than two dimensions (up to seven) are allowed but arrays of more than a few, say four, dimensions are unusual and quickly make large demands on storage capacity.

Arrays of the same "shape" can be grouped in declarations using the DIMENSION alternative, for example

```
REAL,DIMENSION(5,4,3) :: ARRAY_1, ARRAY_2, ARRAY_3
```

reserves 180 REAL locations in the computer for storing the elements of arrays ARRAY_1, ARRAY_2 and ARRAY_3.

Lists of CHARACTER variables, such as the names of players in a football team, could be declared by

```
CHARACTER (LEN=20), DIMENSION (1:11) :: TEAM_NAMES
```

given that the game is soccer (11-a-side) and that no team member has a name more than 20 characters long.

Array terminology

The "rank" of an array is the number of dimensions it has—one for VECTOR, two for ARRAY, three for ARRAY_1 in the above examples. A scalar has a rank of zero. The "shape" of an array is a vector (array of rank one) consisting of the number of elements in each dimension. For example ARRAY given above has shape (5,20), and ARRAY_1 has shape (5,4,3). The "size" of an array is the product of the elements in its shape vector i.e. 100 in the case of our ARRAY example and 60 in the case of ARRAY_1. Arrays are "conformable" if they have the same shape.

Array constructors

Assignment statements for array elements can follow the usual rules, for example

```
A(3,6) = 5.7
```

assigns the value 5.7 to the array element in the third row and sixth column of array A. It is however possible to "construct" the elements of one dimensional arrays using shorthand. Examples of this are

```
V = (/1.0, 2.0, 3.0, 4.0, 5.0/)
```

which is equivalent to

```
V(1) = 1.0;  V(2) = 2.0;  V(3) = 3.0;  V(4) = 4.0;  V(5) = 5.0
```

Array constructor items can themselves be arrays, for example

```
W = (/V,V/)
```

would fill the first ten locations in the array W(I) with elements of V(J) as constructed above.

Moreover, "implied DO loops", see later, can be used in constructors.

Vector subscripts

INTEGER vectors can be used to define subscripts of arrays as in the following example:

```
INTEGER :: SUBSCRIPT (4)
REAL :: FIRST (6), SECOND (4)
SUBSCRIPT = (/1,3,5,6/)
FIRST = (/1.0,2.0,3.0,4.0,5.0,6.0/)
SECOND = FIRST (SUBSCRIPT)
```

The result of this is that

```
SECOND contains FIRST (1), FIRST (3), FIRST (5), FIRST (6)
```

i.e.

```
SECOND contains 1.0,3.0,5.0,6.0
```

Array sections

Parts of arrays or "subarrays" can be referenced by giving a range for one or more of the subscripts, this range being defined by INTEGERs within the bounds of the declared array and separated by a colon. For example if B is a REAL array of dimensions (4 × 5) as shown below:

B(1,1)				B(1,5)
		//////		
		//////		
				B(4,5)

the shaded one-dimensional section can be referenced as B(2:3,3).

If there are no bounds, the colon encompasses all elements in that dimension. For example B(:, 5) refers to all of the 5th column of B.

Note that the array sectioning notation is very similar to that previously used for CHARACTER substrings (Chapter 7). In the soccer team example given above, TEAM_NAMES(1:5) means the five CHARACTER strings representing the names of the first five players. On the other hand, TEAM_NAMES(5)(1:5) means the CHARACTER sub-string representing the first five letters of the fifth player's name.

Array expressions and assignments

Just as arithmetic expressions such as A/B involving simple REAL and INTEGER variables could be written, so A/B has a meaning when A and B are array variables. Such array variables must "conform" in the sense of the array terminology given above, but with this proviso arrays or sections of arrays can be manipulated.

For example if B is (4 × 5) and A is (5 × 4) as below

B(1,1)				B(1,5)
//////	//////	//////		
//////	//////	//////		
B(4,1)				B(4,5)

A(1,1)			A(1,4)
	//////	//////	//////
	//////	//////	//////
A(5,1)			A(5,4)

the shaded (2 × 3) sections could be added, element by element, by writing

    B(2:3, 1:3) + A(3:4, 2:4).

Scalar variables can be used in such expressions and are assumed to be applicable or "broadcast" to all the array elements. Thus if B is (4x5) and X is a scalar, B/X has the effect of dividing all the elements of B by X. Note that if A and B are appropriate arrays A∗B signifies element-by-element multiplication the result of which is *not* the matrix product of A and B (see later).

Array expressions such as those above can then be used in assignment statements for conformable arrays. Thus

```
B = B + 1.
```

results in REAL "one" being added to all the elements of B, whatever its size. The apparent mode mismatch between array B and scalar 1. is covered again by 'broadcasting". As long as arrays A and B are conformable

```
A = A + B
```

replaces the elements of A by the sum of the elements of A and B. This feature of Fortran 90 removes the need to provide array operations such as addition, subtraction, copying, multiplication and division by a scalar, filling an array with scalars (such as zero for "initialisation" purposes) in the form of external SUBROUTINEs in library form (see Chapter 12).

Using sections to avoid DO loops

It will be clear that the sectioning facility can remove the need for DO loops in some instances. For example

```
V(1:4) = V(2:5)
```

is equivalent to, and far neater than,

```
DO I = 1,4
  V(I) = V(I + 1)
END DO
```

It could also help an optimising compiler for specialised, for example parallel, machine architectures.

Care is however needed should iterations in a loop be interdependent. Thus it is not possible simply to replace

```
DO I = 1,4
  V(I + 1) = V(I)
END DO
```

by the single statement

```
V(2:5) = V(1:4)
```

because in the second version no updating of V would occur until all elements of V had been processed.

Thus for example if V is the array 1., 2., 3., 4., 5. the result of the first two assignments above is that V contains 2., 3., 4., 5., 5. whereas in the third version it contains 1., 1., 1., 1., 1. and in the fourth 1., 1., 2., 3., 4.

Array sections need not be of contiguous elements. In the section V(24:36:4) the elements V(24) to V(36) in "strides" of 4 would be selected and if SELECTION is a one dimensional array of size 4 we can write

```
SELECTION = V(24:36:4)
```

Some sorting features can be achieved in this way, for example

```
REVERSE = REVERSE (N:1: −1)
```

would reverse the order of the elements in array REVERSE.

These concepts can be extended to multi-dimensional arrays and lower and upper bounds can be omitted in which case the current bounds will be assumed.

Arrays of "zero" size

If a lower bound of an array exceeds the corresponding upper bound, the array is deemed to have "zero" size and assignments cannot be made to that section. This turns out to be useful in various algorithms involving counting of array elements, when special coding in FORTRAN 77 was necessary to avoid lower bounds exceeding upper bounds (see Chapter 10). Thus if one dimensional array V is declared by

```
REAL,DIMENSION (5) : : V
```

then the section V(6:5) has zero size. Note however that lower and upper bounds must be respected in that sections V(0:2) or V(5:6) are illegal. Another cautionary point is that zero-sized arrays must be conformable if operations on, or assignments to, two or more of them are made.

This is a rather obscure point for average programmers but if an array is declared as R(0) the assignment R = V(6:5) is valid. However, although A(0,2) and B(2,0) are both of zero size, the statement A = B is not valid.

"Dynamic" arrays

Possibly the greatest failing of FORTRAN 77 as an engineering/scientific language was that array sizes had to be known and declared before the program using these arrays was compiled. However, in very many instances, the necessary sizes of arrays only become known after some calculations have been made, or some data input to the program. Arrays whose sizes only become known in the course of a computation are known as "dynamic" arrays, and this facility has been available for many years in other languages (for example ALGOL 60) making FORTRAN 77 very cumbersome by comparison.

Fortran 90 remedies this defect by allowing arrays to be attributed as ALLOCATABLE, illustrated by the following example program:

```
PROGRAM DYNAMIC
IMPLICIT NONE
  ! declare variable space for array A
    REAL,DIMENSION (:,:), ALLOCATABLE: :A
  ! now read in bounds for array A
    READ*, M,N
  ! allocate actual space for array A
    ALLOCATE (A(M,N))
    READ*,A
    PRINT*,2.*SQRT(A)+3.
    DEALLOCATE (A) ! A no longer needed
END PROGRAM DYNAMIC
```

As we have seen, the word "DIMENSION" is optional and often rather laboured in Fortran 90, so that the statement in the above program allocating A can be more concisely written:

```
REAL,ALLOCATABLE: :A(:,:)
```

In the above example the DEALLOCATE statement is of little consequence, but in large programs, storage can be more efficiently handled, especially in a multi-programming environment, if arrays are dispensed with when they are no longer required by the use of DEALLOCATE.

There is an array intrinsic procedure (see Chapter 10) which enables one to test whether or not an array has been ALLOCATED. The LOGICAL expression ALLOCATED(A) will return the value .TRUE. if array A has been ALLOCATED.

The WHERE statement

Array elements can be operated on selectively by means of the WHERE statement which is typified by the following example in which KV is a array:

```
WHERE (KV>.0) KV = −KV
```

which would change the signs of the positive elements in KV. A block structure similar to the block IF construct can be used, in which a sequence of statements is executed on array variables depending on the tested logical array expression. For example

```
WHERE (KV>.0)
   KV = −KV
   A = KV
END WHERE
```

would result in the appropriate elements of A (assuming it is conformable) taking the negatives of KV when these were originally positive.

Again following the IF analogy, there is an IF...THEN...ELSE equivalent construct for arrays which takes the form

```
WHERE (KV>.0)
   Various array assignments
   ELSE WHERE
   Various other array assignments
END WHERE
```

in which the first array assignments will be performed on those elements of the arrays for which the corresponding KV elements are positive while the second array assignments will be performed on the elements of the arrays for which the corresponding KV elements are not positive. Note that scalars cannot be used in this context.

Array subscripts

We have tended to use simple INTEGERs to illustrate array subscripts in examples like A(M,N). It is worth emphasising that array subscripts can be any valid INTEGER expression so that

```
TN (J,(I−1)*NODOF+J) = FUN (IW+I−J+1,J)
```

is a typical array assignment statement.

Implied DO loops

Fortran 90 implementations are supposed to allow arrays to be input or output by a simple statement. Thus if A is an array variable

```
READ*,A
```

or

```
PRINT*,A
```

will read in A in free format, by columns, or print it out. If this facility were not available, a shorthand could be used to avoid the use of DO loops in these circumstances. If V has been declared as an array having dimensions (1:100) and N is less than or equal to 100, all the N elements of V can be read in by the single statement

```
READ*, (V(I),I=1,N)
```

For a two-dimensional array, say A, the equivalent statement would be

```
READ*, ((A(I,J),I=1,M),J=1,N)
```

where M and N are less than the first upper bound and second upper bound respectively of the declared array A. This would read column-wise.

As was mentioned earlier implied DO loops can be used in array constructors, for example

```
INTEGERS_SQUARED = (/(I**2, I=1,N)/)
```

Example: raising a (square) matrix to an INTEGER power

Legendre's method leads to computation times proportional to $\log_2 N$ where N is the size of the matrix. In this method, to calculate A^{2p} where A is a square matrix, if we know A^p we multiply it by itself. Using auxiliary matrices C and D (although computationally we shall only ever need C) we can initialise

$C = I$ (the unit matrix)
$D = A$

The algorithm proceeds as follows:

- if N is even, divide N by 2 and square D
- if N is odd, subtract 1 from N and multiply C by D.

The operation is repeated until N is zero, at which time C contains A^N. If $N = 100$, it will be found that nine multiplications are necessary while if $N = 1000$ the number of multiplications only increases to 15. The program given below will accomplish this task:

```
PROGRAM MATRIX_POWER
! raise a matrix to an integer power
IMPLICIT NONE
REAL, ALLOCATABLE: :A(:,:),C(:,:)
INTEGER: :I,N,M
READ*,N,M
ALLOCATE (A(N,N),C(N,N))
READ*,A;C=0.0
DO I=1,N
  C(I,I)=1.0
END DO
DO
  IF(MOD(M,2)==0) THEN
    M=M/2
    A=MATMUL (A,A)
  ELSE
    C=MATMUL (C,A)
    M=M-1
  IF(M==0) EXIT
  END IF
END DO
PRINT*,C
END PROGRAM MATRIX_POWER
```

As an example, with $N = 4$ and a power M of 5,

$$
\begin{bmatrix} 1 & 0 & 2 & 1 \\ 0 & -3 & 0 & -5 \\ 2 & -5 & 0 & 2 \\ 0 & 9 & 0 & 0 \end{bmatrix}^5 = \begin{bmatrix} 65 & -1425 & 58 & 255 \\ 0 & -13608 & 0 & -4455 \\ 58 & -157 & 36 & 8993 \\ 0 & 8019 & 0 & -10935 \end{bmatrix}
$$

The above example uses the intrinsic procedure MATMUL to multiply two matrices. The full scope of such array intrinsics is given in Chapter 10.

Differences from FORTRAN 77

As we have mentioned before the treatment of arrays in Fortran 90 is possibly the most radical departure from, and improvement on, that permissible in FORTRAN 77. The full declaration statement we have seen in Fortran 90 is of the form

```
REAL,DIMENSION (20)::A,B,C
```

whereas a FORTRAN 77 programmer would have written

```
DIMENSION A(20), B(20), C(20)
```

or, better

```
REAL A(20), B(20), C(20)
```

or, even better

```
PARAMETER (IA=20)
REAL A(IA), B(IA), C(IA)
```

Array sectioning was not permitted in FORTRAN 77, nor were operations on arrays like addition and subtraction. These had to be done instead by external SUBROUTINEs, probably from a SUBROUTINE library.

FORTRAN 77 did not permit zero length arrays.

Dynamic arrays were not a feature of FORTRAN 77. Therefore the equivalent of PROGRAM DYNAMIC listed above would have been something like

```
      PROGRAM QUASI
C ALLOCATE ARRAY SPACE BY PARAMETER STATEMENTS
      PARAMETER (IM=20,IN=10)
      REAL A(IM,IN)
C BE CAREFUL : M CANNOT EXCEED 20 NOR N 10
      READ*,M,N
      READ*,((A(I,J),I=1,M) J=1,N)
      PRINT*,((A(I,J),I=1,M), J=1,N)
      STOP
      END
```

Thus only part of the declared storage space for A would usually be used as shown below:

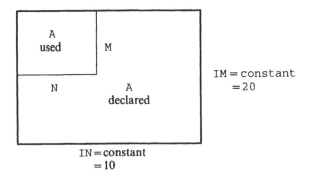

In a "static" version of the above program, the bounds IM and IN would have to be changed by changing the PARAMETER statement before each run of the program.

The advantages of dynamic array storage are obvious and will become even more apparent in Chapter 10 when arrays are used as parameters in FUNCTIONs and SUBROUTINEs.

There was no WHERE statement in FORTRAN 77.

Exercises

1. Write a program which calculates all the prime numbers less than 400, by the method known as the "Sieve of Eratosthenes". In this method we start with a list of integers. The first prime number is known to be 2. All multiples of 2 are therefore removed from consideration. The next prime number is the next highest integer remaining in the list, i.e. 3. All of its multiples are then sieved out. This process is repeated until the next prime number is greater than the square root of 400. The remaining integers will all be prime.

   Print out the prime numbers, and how many of them there are.

   Hint: Store all possible primes in an array.

2. Solve Exercise 1 without storing all the possible primes.

3. Repeat Exercise 1 in Chapter 5 using an array to hold the numbers x_i.

4. Write a program which converts an amount in U.S. dollars or U.K. sterling to the nearest equivalent in another currency, and prints the number of notes and coins of the specified currency required to make up this amount.

   For example, if we consider conversion to the French franc, then the conversion required is to the nearest centime. The relevant data may be prepared as follows:

Record no.	Contents	Meaning
1	FRANCS	name of currency
2	12	number of denominations
3	0.01	1st denomination
4	0.05	2nd denomination
5	0.10	3rd denomination
6	0.20	4th denomination
7	0.50	5th denomination
8	1.00	6th denomination
9	5.00	7th denomination
10	10.00	8th denomination
11	20.00	9th denomination
12	50.00	10th denomination
13	100.00	11th denomination
14	500.00	12th denomination
15	11.9031	rate of exchange (i.e. amount of Francs to the $ or £)
16	3	number of amounts to be converted
17	64.41	1st amount to be converted
18	1.00	2nd amount to be converted
19	25.00	3rd amount to be converted

The output may be printed in the following manner:

```
*********************************************************************
                CONVERSION OF STERLING TO FRANCS
*********************************************************************
                     EXCHANGE RATE = 11.9031
*********************************************************************
```

64.41	STERLING	IS	766.68	FRANCS
DENOMINATION	500.00	100.00	50.00	20.00
NO. REQUIRED	1	2	1	0
DENOMINATION	10.00	5.00	1.00	0.50
NO. REQUIRED	1	0	0	0
DENOMINATION	0.20	0.10	0.05	0.01
NO. REQUIRED	2	0	1	3

```
*********************************************************************
```

5. Stability of a floating pontoon

the Pontoon is square in plan
and volume is fixed.

$$V_p = B*B*D$$

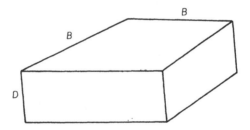

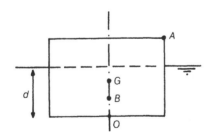

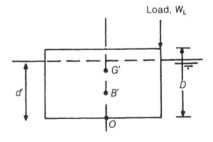

Weight of pontoon, W_p acts
at $G. OG = D/2$
$$OB = d/2$$

Displaced volume, V.
$$= W_p/\rho g$$

Total weight was $W_T (= W_p + W_L)$
acting at G'
$$OG' = ((W_p*OG) + (W_L*D))/W_T$$

$$OB' = d'/2$$
$$V' = W_p/\rho g$$

Metacentric height,
$$MG = I/V - (OG - OB)$$

The load at A is equivalent
to load at centre of deck plus
moment M_L.

$$M_L = W_L*(B/2)$$
$$MG' = I/V' - (OG' - OB')$$

Effect of load is to tilt
the pontoon

Heel angle, $\theta = \dfrac{M_L}{W_T*MG'}$

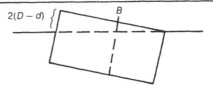

Maximum allowable heel before
deck becomes awash is
$$\tan(\theta_{max}) = (D - d')/(B/2)$$

The problem is to produce a table of heel values for a range of D/B values (0.1, 0.3, 0.5, 0.7, 0.9) and a range of loads (0.1 MN, 0.2, 0.3, 0.4, 0.5). If the heel angle is greater than the maximum then a symbol to indicate "deck awash" replaces the heel angle in the table. If MG' is negative a symbol to indicate "pontoon unstable" is put in the table.

Volume of pontoon V_p is $500\,\text{m}^3$
Weight of pontoon W_p is $2.4525\,\text{MN}$
(Students are advised to work in newtons, kg, m)

Note: For square pontoon

$$\text{draft} = \text{displaced vol.}/B^2$$

$$\text{gives} \quad B = \sqrt[3]{\frac{VB}{D}}$$

$$I = B^4/12$$

Suggested structure

D/B ratio loop   Work out D and B values

Load loop

put appropriate   Work out draft and max heel.
θ value *or* code   Work out MG' and CHECK for stability.
number into a 2D array   Work out heel and CHECK for water on deck.

Print θ array putting in θ values (one decimal place) or the appropriate letter 'A' or 'U' according to the code in the array. This is quite difficult. Substitute a numerical symbol if you prefer.

6. An optional course structure is to be introduced for Civil Engineering students. Only Mathematics is compulsory for all students, their remaining two subjects being chosen from options in Geology, Surveying and Design. Unfortunately, markers are unfair in their estimation of students' abilities. Some give generally higher grades than others. A computer adjustment system is proposed.

Table 8.1 shows the marks gained by a class of 11 students. Write a program which somehow adjusts the marks in the option subjects to a common basis.

Points to consider

1. The tabular form of marks listing encourages the use of an array for storing the marks.
2. All students do Mathematics, so this should serve as a "control".
3. Numbers of students doing options can vary.
4. Some way must be found to indicate that a student did not take a particular course. A student can score zero in a subject, but not a negative number.
5. The first adjustment to be contemplated is to the mean or average mark in each option.
6. If you have time, go on to consider adjusting the "spread" of marks.
7. Write a short description of the scheme you propose, with a structure chart, before doing any coding.

Table 8.1

		Marks Gained in		
Subject	1	2	3	4
Student	Mathematics	Geology	Surveying	Design
1	56	38		56
2	34	27	60	
3	72	55		48
4	11		72	59
5	43		50	45
6	80	62		50
7	22		55	49
8	95	30	65	
9	0	40	51	
10	37		49	49
11	21	50		50

7. Write a program which will read in a list of REAL numbers, sort them into descending order, and print them out. Note: See Chapter 11 for better methods of sorting.

9

Yet More on Input/Output: FILES

Introduction

In large programs information can be received from, and output to, various media or devices such as discs, tapes, cartridges and so on as well as the keyboard and screen we have discussed until now. A collection of data on any of these devices is called a "file" and is composed of a series of "records". For example if an output file appears on a screen, a record is likely to be one line on the screen.

Unit numbers

Devices from which data is to be read, or to which data is to be output, can be allocated a "unit number" which is a positive INTEGER or scalar INTEGER-valued expression. On most systems, the keyboard and screen will have "default" unit numbers—5 for the keyboard and 6 for the screen for example. Using these unit numbers, the list directed READ and PRINT statements we have used so far become a little different, the unit numbers appearing in parentheses after READ and the PRINT statement being replaced by WRITE.
    For example

```
READ*,A,B,C
```

becomes

```
READ(5,*)A,B,C
```

and is exactly equivalent if unit 5 is the default, allowing free format input which we recommend. The longer version READ(*,*) is equivalent to READ*, while WRITE(*,*) is equivalent to PRINT*. Indeed these longer forms are actually themselves shorthand for the complete versions READ (UNIT = *, FMT = *) and WRITE (UNIT = *, FMT = *) respectively.

We can override the defaults so that input and output can be received from and sent to various devices, or to various locations within a given device such as a disc. For example we could initialise input and output devices by writing

```
OPEN (UNIT=10,FILE='P53A.DAT', STATUS='OLD')
OPEN (11,FILE='P53A.RES')
```

where we can see that "UNIT=" is optional as long as it appears first in the list of options. The first of these statements would allow us to read unformatted data from an existing file called P53A.DAT using statements like

```
READ(10,*)A,B,C
```

To output to the file defined above as P53A.RES we might use a statement like

```
WRITE(11,'(5E12.4)') A(1:5)
```

which produces formatted output and is shorthand for

```
WRITE(UNIT=11, FMT='(5E12.4)')A(1:5)
```

For "FMT=" to be omitted it must be the second entry in the input or output list.

Non-advancing I/O

We have seen previously that each call to an input or output statement like READ or PRINT causes a complete record to be processed. It can be convenient if the file position after READ or PRINT (WRITE) is not advanced to be ready to read from (write to) a fresh record. This facility in Fortran 90 is called "non-advancing" I/O, and is achieved by including the character string 'NO' with the optional ADVANCE = specifier as follows

```
READ(10,'(F3.0)',ADVANCE='NO') A,B,C
```

It cannot be used for list-directed I/O.

Unformatted output

We have recommended the invariable use of unformatted input and if output is not to be inspected, but for example used as input to another program on the same or another computer with the same internal configuration, we recommend unformatted output as well. This avoids chances of erroneous interpretation of FORMAT, roundoff and so on. To do this, the FMT specifier is simply omitted from the WRITE statement, for example

```
WRITE(11) A(1:5)
```

Manipulating FILES

We have so far seen how to make a FILE available to a Fortran 90 program by the use of the OPEN statement. This is just one of a number of FILE manipulation statements which we can collect under three headings: FILE positioning statements, FILE initiation and closure statements and FILE status enquiry statements.

(i) FILE positioning

It is quite common programming practice to output a record or series of records to a FILE and then need to read these records back into the program again at a later stage. To do this, a typical statement would be

```
BACKSPACE 10
```

where "10" is the unit number of the FILE being manipulated. The effect of this statement is to move the reading (writing) position to the beginning of the current record if positioned within that record and to the beginning of the previous record if positioned between records. If the current reading (writing) position is at the beginning of a FILE, then BACKSPACE has no effect.

Since BACKSPACE only moves the reading (writing) position by one record, it is useful to be able to move to the beginning of a complete FILE. The statement to enable this to be done is typically

```
REWIND 8
```

A computer system will usually keep track of the end of a FILE by marking it with an end of FILE marker, but in case of potential error, a marker can be explicitly placed at the end of a FILE by writing, for example

```
ENDFILE 7
```

(ii) FILE initiation and closure

We have already seen an example of an OPEN statement for initiating a FILE and to complement it there is an equivalent closure statement, CLOSE.

The full form of the OPEN statement is

```
OPEN(UNIT = unit number, options list)
```

where, as we have seen, "UNIT =" is optional. The "options list" is potentially very comprehensive, as shown in Appendix 4, but will typically contain only a filename and possibly a STATUS as in our earlier example. Another useful option is ACTION which encourages programmers to be clear whether they are reading from a file, or writing to it, or both.

To complement the OPEN statement there is an equivalent CLOSE statement. All FILEs are automatically closed on the completion of a program but there are some optional actions particularly to do with the STATUS of disconnected FILEs. The full CLOSE statement takes the form

```
CLOSE(UNIT=unit number, options list)
```

where as usual "UNIT =" is optional. The list of options is much smaller than for OPEN and consists of IOSTAT = **ios** with the same meaning as in OPEN, ERR = **error label** with the same meaning as in OPEN, and STATUS = **st** where **st** is a CHARACTER expression having one of the two values 'KEEP' or 'DELETE'. If the first of these is specified, a FILE continues to exist and can be reconnected later. If 'DELETE' is specified, the FILE is lost after the CLOSE statement is executed. A FILE opened with STATUS = 'SCRATCH' cannot be saved using 'KEEP', but with this exception, the default is STATUS = 'KEEP'. An example of a CLOSE statement might be

```
CLOSE(14,IOSTAT=IFAIL, STATUS='DELETE')
```

(iii) Enquiry about FILE STATUS

At any time during the execution of a program it is possible to make enquiries about the status of a file or a unit by means of the INQUIRE statement.

There are three versions, known as INQUIRE by file, INQUIRE by unit and INQUIRE by output list. The first two have similar specification lists and take the respective forms:

```
INQUIRE (UNIT = u, options list)
```

where as usual "UNIT =" is optional, and

```
INQUIRE (FILE=filename, options list)
```

The options list specifiers are given in Appendix 4.

Exercises

1. Two data files contain lists of students identified by a registration number, followed by the student's name and average examination score. The files are sorted in order of increasing registration number. Write a program to merge the two files into a third file also sorted in order of increasing registration number. Test your program on the following files:

12157	STEVEN L. ARCHER	68.5
12169	ANDREW J.N. BALLINGER	43.2
13012	KONSTANTINOS BARMPAS	60.2
13565	DANIEL BARTLE	55.3
13829	TASVEER A. BEG	80.5
13900	ALASTAIR J. BUSH	44.4
13921	IAN T. CHEONG	69.2

and

12160	QUENTIN P. HOLLAND	53.3
12199	MALCOLM HORNE	60.5
13011	NICOLAS D. HUDSON	45.5
13229	JAMES A.L. HYSLOP	59.1
13626	AARON JOHNSON	43.2
13800	FRANCIS JOSEPH	81.5
13905	ANTONY R. LUCAS	60.0
13906	MATTHEW D. NEWMAN	88.8
13999	BENJAMIN W. HEYS	75.6

10

Subprograms: FUNCTIONS and SUBROUTINES

> *Glendower: I can call up spirits from the vasty deep*
> *Hotspur: But will they come when you call for them?*
> Shakespeare

Introduction

In Chapter 4 we introduced the idea of subprograms. We saw that it makes sense in terms of program construction to build programs from a series of constituent "building blocks" called subprograms or "procedures" and indeed that something like 100 of these (mainly FUNCTIONs) are perceived to be of such widespread utility that they are provided as part of the language. These are the intrinsic procedures listed in full in Appendix 3.

In Chapter 12 we shall see that access to useful code can be considerably expanded by the use of "libraries" of subprograms, some of which may be commercial products, which are completely independent of, or "external" to, a particular user program.

Libraries of external subprograms can also be user-produced, but some subprograms can be "contained" within programs or other subprograms. In this chapter we learn how to write subprograms, remembering that the essential difference between a FUNCTION and a SUBROUTINE is that the former "returns" a single quantity (of any TYPE and even an array) to the program which "calls" it whereas the latter returns several quantities, usually of several TYPEs.

Main program and subprograms

In this chapter we limit discussion to the case where there is a single "main" program which calls internal subprograms. An extension of this idea, in which subprograms can be nested in the form of MODULEs is left to Chapter 11. With this simplification, a typical program structure is as shown below:

```
┌────────────────────────────────────┐
│           Main Program             │
│                                    │
│   ┌──────────────────────────┐     │
│   │   Internal Subprograms    │     │
│   └──────────────────────────┘     │
│                                    │
│        End of Main Program         │
└────────────────────────────────────┘
```

We have already seen many examples of main programs which have the recommended form:

```
PROGRAM name
    Fortran 90 statements
END PROGRAM name
```

although, strictly speaking, only END is essential after the Fortran 90 statements.

To augment the main program by internal subprograms (as in the diagram above) the form becomes

```
PROGRAM name
    Fortran 90 statements
CONTAINS
    Internal subprograms (FUNCTIONs and/or SUBROUTINEs in Fortran 90)
END PROGRAM name
```

The first type of subprogram has the following form:

```
FUNCTION name (arguments) RESULT (name of result)
    Fortran 90 statements
    name of result = Fortran 90 assignment
END FUNCTION name
```

For example the intrinsic FUNCTION named SIN which we saw in Chapter 4 has the form:

```
FUNCTION SIN (ANGLE) RESULT (SIN_RESULT)
    IMPLICIT NONE
    REAL : : ANGLE, SIN_RESULT
    Fortran 90 statements
    SIN_RESULT = Fortran 90 assignment
END FUNCTION SIN
```

In this case there is a single argument, namely the value of the angle (in radians) whose sine is required. However there can be many arguments, as long as the result is a single quantity to which an assignment can be made. The FUNCTION name can then be freely used in programs by statements like

```
A = SIN(X) * SIN(Y)
```

We see that "ANGLE" is a "dummy" argument which is replaced by actual arguments such as "X" and "Y" when the program is run.

The concept of FUNCTIONs can be further generalised to include the possibility of the FUNCTION calling itself. Such "recursive" use of FUNCTIONs is described at the end of this chapter.

Recursive FUNCTIONs are the reason for the use of the RESULT "clause". The above form is rather cumbersome for simple FUNCTIONs and we could just as well write

```
FUNCTION SIN(ANGLE)
   IMPLICIT NONE
   REAL::ANGLE
   Fortran 90 Statements
   SIN = Fortran 90 assignment
END FUNCTION SIN
```

Since the FUNCTION name is a Fortran 90 name, and we advocate "strong" TYPEing we should really have written REAL FUNCTION SIN (ANGLE). Other TYPEs of FUNCTION, for example INTEGER, LOGICAL, COMPLEX, CHARACTER can be written.

The second type of subprogram has the following form:

```
SUBROUTINE name (input arguments, output arguments)
   Fortran 90 statements
   output arguments = Fortran 90 assignments
RETURN
END SUBROUTINE name
```

As a (not useful) example, suppose we wanted to construct a SUBROUTINE structure to calculate the reciprocals of two real numbers. We might write

```
SUBROUTINE RECIPROCAL (A,B,C,D)
   IMPLICIT NONE
   REAL : : A,B,C,D
   C=1./A
   D=1./B
   RETURN
END SUBROUTINE RECIPROCAL
```

and embed this in a main program as follows:

```
PROGRAM MAIN
   IMPLICIT NONE
   REAL : : R1,R2,R3,R4,R5,R6,R7,R8
     READ*, R1,R2,R3,R4
       CALL RECIPROCAL (R1,R2,R5,R6)
       CALL RECIPROCAL (R3,R4,R7,R8)
   PRINT*, R5,R6,R7,R8
CONTAINS
   SUBROUTINE RECIPROCAL (A,B,C,D)
   IMPLICIT NONE
     REAL : : A,B,C,D
       C=1./A;D=1./B
     RETURN
   END SUBROUTINE RECIPROCAL
END PROGRAM MAIN
```

Note that to use the SUBROUTINE named RECIPROCAL in the main program we must precede it by the Fortran 90 keyword CALL. The RETURN statement "returns" program control to the main program on completion of the SUBROUTINE call. It is optional, rather like STOP at the end of a main program, but may serve a useful purpose in reminding the programmer of the point of return in the main program.

Arguments A, B, C and D are "dummies" and are replaced by actual arguments R1 etc at the times the calls are made. The program would read in four REAL numbers and print out their reciprocals.

Input and output arguments

However unlikely the above example is as a serious program it illustrates the basic feature of SUBROUTINE subprograms in that they generally have some arguments intended to be input parameters and other arguments intended to be output parameters. A third possibility, to be used with care, is when an argument changes from input to output in the course of the execution of the SUBROUTINE.

It is good programming discipline to be clear about the intention with which arguments are used. Fortran 90 encourages this, by the provision of three specifications: INTENT(IN), INTENT(OUT) and INTENT(INOUT). In the third case the argument would be changed by the SUBROUTINE. We could modify RECIPROCAL as follows using this facility:

```
SUBROUTINE RECIPROCAL (A,B,C,D)
  IMPLICIT NONE
  REAL,INTENT(IN)::A,B
  REAL,INTENT(OUT)::C,D
    C=1./A ; D=1./B
  RETURN
END SUBROUTINE RECIPROCAL
```

Arrays as arguments in subprograms

A major deficiency in FORTRAN 77 (see Differences from FORTRAN 77 below) arose from the lack of a dynamic array structure which we discussed in Chapter 8. This had a serious effect on the use of arrays as arguments in FORTRAN 77 SUBROUTINEs, because these arrays had to have a fixed size when the SUB-ROUTINE was compiled. Extra arguments were therefore necessary to take care of the nominal, fixed, sizes of the arrays to be used as well as their actual sizes when the program was run. The provision of a "dynamic" array structure in Fortran 90 removes these restrictions, and arrays can be passed as arguments with at most only their actual run-time sizes as extra arguments and, by the use of array intrinsics (see below), not even these.

For example, suppose we wish to write a SUBROUTINE to multiply two matrices together, row by column, in the usual manner. In Fortran 90 we might write:

```
      SUBROUTINE MATRIX_MULT(A,B,Y,L,M,N)
! multiply A(L,M) by B(M,N) to give Y(L,N)
      IMPLICIT NONE
      INTEGER::I,J,K,L,M,N
      REAL,INTENT(IN),DIMENSION(L,M)::A
      REAL,INTENT(IN),DIMENSION(M,N)::B
      REAL,INTENT(OUT),DIMENSION(L,N)::Y
      Y=0.0
        DO K=1,N
          DO I=1,M
            DO J=1,L
              Y(J,K)=Y(J,K)+A(J,I)*B(I,K)
            END DO
          END DO
        END DO
      END SUBROUTINE MATRIX_MULT
```

In principle, since the output of this SUBROUTINE is a single quantity of an array of elements of REAL TYPE it could be arranged as a FUNCTION (see below).

As we have remarked before, the DIMENSION attribute is rather laboured and we could replace the above statements containing it by, for example

```
REAL,INTENT (IN) :: A(L,M)
```

After the next section, we shall see that the intrinsic procedures provided in the language for arrays enable us to simplify the above matrix multiplication routine considerably.

Intrinsic procedures involving array computations

In FORTRAN 77, it was necessary to have SUBROUTINEs available for all array computations such as array addition, multiplication of an array by a scalar and so on. Many of these are redundant in Fortran 90 since, as we saw in Chapter 8, arithmetic operations are defined to apply to arrays as well as to scalars. Thus if A,B and C are arrays and S is a scalar, we can write

```
C=A+B
C=S*A
A=0.0
```

and so on, as long as A,B and C satisfy conformability requirements. Scalars like 0.0 and S are "broadcast" to all the array elements.

The above features are supplemented in Fortran 90 by a few useful intrinsic FUNCTIONs which operate on arrays. Indeed the matrix multiplication cast above as a SUBROUTINE is one of these and takes the form

```
FUNCTION MATMUL(A,B)
```

As long as A,B and C satisfy conformability requirements we can write

```
C = MATMUL(A, B)
```

which is also valid as a matrix by vector multiplication where C and B are conforming vectors.

If A and B are conforming vectors, their dot product can be found by

```
FUNCTION DOT_PRODUCT(A, B)
```

and assigned to a scalar by, for example,

```
S = DOT_PRODUCT(A, B)
```

A third useful FUNCTION can be used to form the transpose of a two-dimensional array. It takes the form

```
FUNCTION TRANSPOSE(A)
```

and if A and AT have the right shapes, one can write

```
AT = TRANSPOSE(A)
```

Intrinsic procedures for inspecting arrays

In addition to the above array computation procedures, several FUNCTIONs are available for inspecting arrays. A complete list is given in Appendix 4, and amongst the more useful are:

```
FUNCTION MAXVAL(A)  !         Returns the element of an array A of
                    !         maximum value
FUNCTION MINVAL(A)  !         Returns the element of an array A of
                    !         minimum value
FUNCTION MAXLOC(A)  !         Returns the location of the maximum element
                    !         of array A
FUNCTION MINLOC(A)  !         Returns the location of the minimum element
                    !         of array A
FUNCTION PRODUCT(A) !         Returns the product of all the elements of A
FUNCTION SUM(A)     !         Returns the sum of all the elements of A
FUNCTION LBOUND(A,1)!         Returns the first lower bound of A etc
FUNCTION UBOUND(A,1)!         Returns the first upper bound of A etc
```

Masking

The first six of the above procedures allow an optional extra argument called a "masking" argument. For example the statement

```
ASUM = SUM(COLUMN, MASK = COLUMN > = .0)
```

will result in SUM being the sum of the positive elements of the one-dimensional array COLUMN. "MASK =" is optional. Note that since SUM is a procedure name its use as a simple variable is discouraged. Indeed compilers will not allow the use of SUM as a simple variable in a program in which the intrinsic procedure SUM is called.

Useful intrinsic procedures whose only argument is a "mask" are called ALL, ANY and COUNT. The statement

```
CANDIDATES = COUNT (COLUMN > = 0)
```

will result in INTEGER CANDIDATES containing the number of positive entries in the one-dimensional array COLUMN.

The dimension argument

For multidimensional arrays it is useful to be able to perform operations such as SUM on a particular dimension of an array. Therefore, in addition to LBOUND and UBOUND given above, procedures ALL, ANY, COUNT, MAXVAL, MINVAL, PRODUCT, SIZE (see below) and SUM can have an optional extra argument (see Appendix 3) allowing a particular dimension of the array to be specified. Thus

```
BANDWIDTH = MAXVAL (G, 1, G > 0) — MINVAL(G, 1, G > 0)
```

will calculate the maximum difference between positive elements along the first dimension of G.

As an illustration of array intrinsics, consider the following program:

```
PROGRAM TEST_ARRAY_INTRINSICS
! test of all, any etc
IMPLICIT NONE
INTEGER :: N; REAL, ALLOCATABLE :: A(:)
READ*, N
ALLOCATE (A(N)); READ*, A
PRINT*, ALLOCATED (A)
PRINT*, ALL (A > .0), ANY (A < .0)
PRINT*, COUNT (MASK = A > .0), COUNT (MASK = A < .0)
PRINT*, MAXLOC (A), MAXVAL (A)
PRINT*, MINLOC (A), MINVAL (A)
PRINT*, PRODUCT (A), SUM (A)
PRINT*, LBOUND (A), UBOUND (A)
END PROGRAM TEST_ARRAY_INTRINSICS
```

If A is the array 1.0 2.0 3.0 4.0 the result will be

```
.TRUE.
.TRUE. .FALSE.
4     0
4     4.0
1     1.0
24.0  10.0
1     4
```

Altering arrays

The "shape" of an array is an INTEGER vector (rank 1 array) containing the "extents" of each of the subscripted variables. For example the array BOX declared as

```
REAL, DIMENSION (−5:5, −3:3, −2:2) :: BOX
```

has the shape (11,7,5). An intrinsic FUNCTION called SHAPE is provided which
returns this shape as an INTEGER vector, for example

```
ARRAY_SHAPE = SHAPE (BOX)
```

The "size" of an array is the total space it occupies in the computer
$-11 \times 7 \times 5 = 385$ in the case of BOX above. The intrinsic function called SIZE
returns this size as a simple INTEGER, for example

```
I = SIZE (BOX)
```

would result in I taking the value 385. As explained above, the extents of individual
dimensions can also be interrogated, for example

```
J = SIZE (BOX,2)
```

would result in J taking the value 7.

It is possible to "reshape" arrays. For example, suppose we wish to reshape BOX
which has the shape ARRAY_SHAPE into a one dimensional array (vector) LINE. Of
course the sizes of BOX and LINE would be the same but LINE may have been declared as

```
REAL :: LINE (385)
```

In that case we can say

```
LINE = RESHAPE (SOURCE = BOX, SHAPE = SHAPE (LINE))
```

If LINE and BOX are of different sizes, this is a more risky operation involving
truncation or padding. As usual "SOURCE =" and "SHAPE =" are optional.

A similar effect can be achieved by intrinsic FUNCTION PACK (there is an
equivalent UNPACK). Finally FUNCTION SPREAD enables copies of arrays or
sections of them to be made. The following small program illustrates the use of array
altering FUNCTIONs.

```
PROGRAM TEST_SPREAD_AND_PACK
! test the intrinsics to alter arrays
IMPLICIT NONE
REAL :: A(3,4), B(12), C(12), D(2,3,4)
READ*,A
PRINT*, SIZE(A), SHAPE(A)
B = RESHAPE (A, SHAPE(B)); PRINT*,B
C = PACK (A,.TRUE.); PRINT*,C
D = SPREAD (A,1,2); PRINT*,D
END PROGRAM TEST_SPREAD_AND_PACK
```

This program copies array A into vectors B and C while D contains two copies of
A with the rank extended as shown.

Using these procedures for examining array sizes, we could rewrite our previous
matrix multiplication subroutine as follows

```
    SUBROUTINE MATRIX_MULT (A,B,Y)
  ! Multiply A by B to give Y
    IMPLICIT NONE
    INTEGER :: I,J,K,L,M,N
    REAL,INTENT (IN) :: A(:,:), B(:,:)
    REAL,INTENT (OUT) :: Y(:,:)
    L=SIZE (A,1); M=SIZE (A,2); N=SIZE (B,2)
    Y=0.0
    DO K=1,N
      DO I=1,M
        DO J=1,L
          Y(J,K)=Y(J,K)+A(J,I)*B(I,K)
        END DO
      END DO
    END DO
    END SUBROUTINE MATRIX_MULT
```

Of course the intrinsic procedure MATMUL should always be used for this task.

Interfaces with procedures

When a subprogram is internal and contained within a main program, checks can be carried out at compile time to see, for example, that the numbers and TYPEs of the arguments are compatible between procedure declarations and calls. This interface between program and subprograms is said to be "explicit".

However, if subprograms exist externally, for example in pre-compiled libraries, there are no means for the compiler to carry out any checks when compiling the host program. In this case the interface between program and subprograms is said to be "implicit".

It is good practice to specify the names of such external subprograms at the beginning of the main program by using the statement

```
    EXTERNAL list of procedure names
```

This in itself does not specify the interface, but it can be done by means of an "interface block" of the form

```
    INTERFACE
      interface body
    END INTERFACE
```

where the interface body usually contains a list of subprogram name declarations with arguments but without any associated code. Thus an explicit interface is formally created which prevents confusion between, for example, external procedures and intrinsic procedures of the same name (which would be ignored). We return to the use of external subprograms in Chapter 12.

FUNCTIONS returning arrays

We have noted several times that in Fortran 90, in contrast to FORTRAN 77, a FUNCTION can return an array result. In the example given below this is somewhat dangerous in that the array argument A is altered during the FUNCTION call. There is also an overhead of an extra array (the returning FUNCTION name "INVERT"). It would be possible to avoid this by returning another quantity, say a flag which monitored the division by "CON" and returned a message should CON be very small. The FUNCTION inverts a square matrix A of size $N \times N$ by the Gauss–Jordan transformation method. Some use is made of array sectioning.

```
PROGRAM INVERT_A_SQUARE_MATRIX
! Gauss-Jordan transformation
IMPLICIT NONE
REAL,ALLOCATABLE::A(:,:), X(:)
INTEGER::N; READ*,N; ALLOCATE (A(N,N),X(N))
READ*,A; READ*,X; PRINT*,A; PRINT*,X
PRINT*,INVERT(A); PRINT*,MATMUL(A,X)
CONTAINS
FUNCTION INVERT(A)
! Gauss-Jordan method
IMPLICIT NONE
REAL::A(:,:)
REAL::INVERT(UBOUND(A,1),UBOUND(A,2))
! local variables
INTEGER::I,K,N;REAL::CON;N=UBOUND(A,1)
DO K=1,N
  CON=A(K,K); A(K,K)=1.
  A(K,:)=A(K,:)/CON
    DO I=1,N
      IF(I/=K) THEN
        CON=A(I,K); A(I,K)=.0
        A(I,:)=A(I,:)-A(K,:)*CON
      END IF
    END DO
END DO
INVERT=A
END FUNCTION INVERT
END PROGRAM INVERT_A_SQUARE_MATRIX
```

Procedures as arguments

So far we have used variables as procedure arguments, but it is perfectly possible to use names of FUNCTIONs as well. Thus one FUNCTION can call another or several others. This is one of the very few instances in which the "specific" names of intrinsic FUNCTIONs we saw briefly in Chapter 4 can matter (as distict from their "generic" names), because the FUNCTION's name predetermines its TYPE.

Optional arguments

One of the reasons given by FORTRAN 77 programmers for their preference for COMMON blocks (see below : Obsolescent Features) is that argument lists "get too long". In Fortran 90 it is possible to declare arguments as OPTIONAL, for example

```
REAL,OPTIONAL :: THIS,THAT
```

and to omit them from procedure calls if desired. Thereafter it may be necessary within a procedure to find whether a certain argument has been specified in a particular call. This is possible using the intrinsic LOGICAL FUNCTION called PRESENT which only returns the value .TRUE. if its argument has been specified. Otherwise it returns the value .FALSE.. For example

```
IF (PRESENT (ARGUMENT_X)) B = ARGUMENT_X**2
```

Another solution to the lengthy argument list problem is through the use of MODULEs, see Chapter 12.

Recursive procedures

By preceding a procedure declaration by the Fortran 90 word RECURSIVE, a procedure can be made to call itself.

Although, as mentioned previously, there seems to be little point for straight-forward FUNCTIONs, it is possible for the result of a FUNCTION evaluation to have a name different from the FUNCTION name. This is illustrated again below, and involves the use of the Fortran 90 keyword RESULT. Suppose we wish to sum up the series $\sum_{i=m}^{n} b^i$ where b is a REAL constant. We can write

```
FUNCTION SUMS (M,N,B) RESULT (SUM_OF_SERIES)
  IMPLICIT NONE
  INTEGER :: M,N,I
  REAL :: SUM_OF_SERIES, B
    SUM_OF_SERIES = 0.0
    DO I = M,N
      SUM_OF_SERIES = SUM_OF_SERIES + B**I
    END DO
END FUNCTION SUMS
```

and then make assignments as usual, such as

```
A = SUMS (10,100,0.1)
```

This facility really comes into play when a FUNCTION is used recursively. In this case, to avoid confusion, the result *must* have a name different from that of the FUNCTION.

A commonly quoted example is the factorial function which we could write as follows:

```
RECURSIVE INTEGER FUNCTION FACTORIAL (N) RESULT (FAC_N)
  IMPLICIT NONE
  INTEGER :: N
    SELECT CASE (N)
      CASE (1)
        FAC_N = 1
      CASE DEFAULT
        FAC_N = N*FACTORIAL (N-1)
    END SELECT
END FUNCTION FACTORIAL
```

There is no corresponding difficulty for recursive SUBROUTINEs.

In the above example, note the "when to stop" condition when $N = 1$. To understand recursion, consider (and program) what would happen without the "when to stop" rule. Also see Exercise 2, Chapter 11 which is about the game of "Hanoi".

Structure chart representation

Our convention for representing procedures is:

> Use Algorithm x

Here, the appropriate algorithm is usually complicated enough to merit its own structure chart and must be cross-referenced with enough input/output information.

Differences from FORTRAN 77

Many of the features described in this chapter were not available in FORTRAN 77. There used to be no means of differentiating between internal and external subprograms, so CONTAINS could not be used. A major difference is that an array could not be the result of a FUNCTION call, and a SUBROUTINE had to be used if the single result of a subprogram was an array.

There was no means of signifying INTENT in the declaration of subprogram arguments.

We have referred at length to the difficulty of passing arrays as arguments in SUBROUTINEs in FORTRAN 77. In previous versions of the language, the SUBROUTINE given earlier for multiplying matrices together would have taken a form something like

```
      SUBROUTINE MATMUL (A,IA,B,IB,Y,IY,L,M,N)
C
C     IA,IB,IC ARE THE FIXED COLUMN SIZES
C     OF A,B,Y.
C     THE RUN-TIME SIZES ARE A(L,M),B(M,N),
C     Y(L,N)
C
      REAL A(IA,*),B(IB,*),Y(IY,*)
        DO 1 I=1,L
          DO 1 J=1,N
            X=0.0
            DO 2 K=1,M
2           X=X+A(I,K)*B(K,J)
          Y(I,J)=X
1       CONTINUE
      RETURN
      END
```

In FORTRAN 77 there were no intrinsic procedures involving arrays, and no means for specifying INTERFACEs with procedures. Optional arguments were not permitted, nor were recursive procedures. It was not possible for a FUNCTION to return a result with a name different from that of the FUNCTION itself.

As an example of the superiority of Fortran 90, the following code from a text on Finite Element analysis (Smith and Griffiths, 1988*) is given below in both dialects:

* Smith, I.M. and Griffiths, D.V. *Programming the Finite Element Method*, 2nd Edition, John Wiley and Sons (1988)

Fortran 90

```
KM=0.0
GAUSS_POINTS : DO I=1,NGP;DO J=1,NGP
          CALL FORMLN (DER,FUN,SAMP,I,J)
          CALL TWOBYTWO (MATMUL(DER,COORD),JAC1, DET)
          DERIV=MATMUL (JAC1,DER)
          BEE=0.0;CALL FORMB (BEE,DERIV)
   KM=KM+MATMUL (MATMUL(TRANSPOSE(BEE),DEE),BEE) &
                            *DET*SAMP(I,2)*SAMP(J,2)
   END DO ; END DO GAUSS_POINTS
```

FORTRAN 77

```
       CALL NULL (KM,IKM,IDOF,IDOF)
       DO 20 I=1,NGP
       DO 20 J=1,NGP
       CALL FORMLN(DER,IDER,FUN,SAMP,ISAMP,I,J)
       CALL MATMUL(DER,IDER,COORD,ICOORD,JAC,IJAC,IT,NOD,IT)
       CALL TWOBYTWO(JAC,IJAC,JAC1,IJAC1,DET)
       CALL MATMUL(JAC1,IJAC1,DER,IDER,DERIV,IDERIV,IT,IT,NOD)
       CALL NULL(BEE,IBEE,IH,IDOF)
       CALL FORMB(BEE,IBEE,DERIV,IDERIV,NOD)
       CALL MATMUL(DEE,IDEE,BEE,IBEE,DBEE,IDBEE,IH,IH,IDOF)
       CALL MATRAN(BT,IBT,BEE,IBEE,IH,IDOF)
       CALL MATMUL(BT,IBT,DBEE,IDBEE,BTDB,IBTDB,IDOF,IH,IDOF)
       QUOT=DET*SAMP(I,2)*SAMP(J,2)
       CALL MSMULT (BTDB,IBTDB,QUOT,IDOF,IDOF)
20     CALL MATADD(KM,IKM,BTDB,IBTDB,IDOF,IDOF)
```

Obsolescent features of FORTRAN 77

The only means described in this chapter for transmitting information into and out of subprograms involved passing the information as arguments in the subprogram calls. Many, particularly large, programs which have been developed over the years in FORTRAN do not use this facility, but instead allow communication between program units by means of "COMMON blocks". These may be "named" as in the statement.

```
COMMON/NAMED/FRED,JIM,HARRY (25)
```

or "blank" as in the statement

```
COMMON//BLANK (50000)
```

Usually "blank COMMON" is used to define a large data area to be used by different subroutines as working space while "named COMMON" statements often appear identically in every subprogram written in this style. This is tedious and prone to error, and we advise that COMMON blocks are unnecessary in Fortran 90.

If identical "named COMMON" statements do not appear in the different program units, the dangerous practice of associating data with different variable names may be being attempted. Thus in the above example, the use of the block called "NAMED" in a different program unit by the statement

```
COMMON/NAMED/TOM,JAMES (20), BILL (6)
```

would associate the data in JAMES (10) with that in HARRY (9) and the data in BILL (1) with that in HARRY (20). It should be self-evident that this is obscure and likely to lead to difficulties.

Similar remarks apply to the storage association statement EQUIVALENCE which is unnecessary in Fortran 90.

Exercises

1. Repeat Exercise 1 in Chapter 5 using intrinsic array procedures (for example SUM).

2. Write a SUBROUTINE which takes three arguments, namely a REAL one dimensional array and two REAL variables, and returns in the two variables the mean and the standard deviation (see Exercise 1 in Chapter 5) of the elements of the array.

3. Write a FUNCTION which takes three arguments, namely two REAL one dimensional arrays and an integer (say n), and returns as its value the correlation coefficient of the first n elements of the two arrays.

   The correlation coefficient of two sets of n values, $x_1 x_2, \ldots, x_n$ and $y_1, y_2, \ldots, y_n$, is

$$\frac{n\sum_{i=1}^{n} x_i y_i - \sum_{i=1}^{n} x_i \sum_{i=1}^{n} y_i}{\sqrt{n\sum_{i=1}^{n} x_i^2 - (\sum_{i=1}^{n} x_i)^2} \sqrt{n\sum_{i=1}^{n} y_i^2 - (\sum_{i=1}^{n} y_i)^2}}$$

4. Using the above subprograms, write a program which:
   (a) reads pairs of values (x_i, y_i);
   (b) computes the means and the standard deviations of the x_i and y_i values respectively;
   (c) checks the reliability of the input data. (There may be bad input data as a result of careless typing or experimental errors. This check looks for any pair of x_i and y_i where either x_i or y_i is more than three times the standard deviation away from the mean of the x_i values or the y_i values respectively. Such a pair is "bad" and is rejected).
   (d) calculates the standard deviations and the correlation coefficient of the "good" x_i and y_i values.

   The output of the program should include the number of pairs of x_i and y_i values read in, the number of "bad" pairs, the means and standard deviations, and the correlation coefficient of the "good" x_i and y_i values.

5. A crude graph may be "drawn" on a terminal by printing suitable characters. The resolution will be poor but the result could be adequate for simple applications.

   A stress–strain curve for a typical textile material is shown below (Hooke's law does not apply!). Such a curve can be represented reasonably well by an equation of the form:

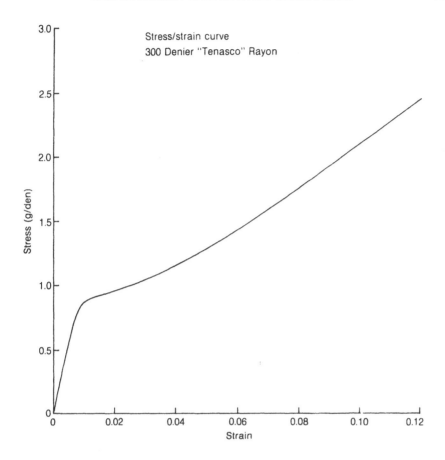

$$y = (ax^n - d)e^{-bx} + cx + d$$

where a, b, c, d and n are numbers which characterise this particular material.

Write a program to print out a graph of y against x for $a = 620.0, b = 130.0, c = 17.0,$ $d = 0.4$ and $n = 1.3$ and compare it with the test result in the figure.

Use dots for the axes and asterisks for the graph points. Print the x-axis vertically down the page (so that a line may be printed without storing previous values).

Hints: (a) Generate the (x, y) coordinates and store them in an array
(b) Find the maximum and minimum y values for scaling purposes.

6. INTEGER solutions of the equation $x^3 + y^3 = a^3 + b^3$ may be found by the following method:

(i) For all possible values of x and y in some range $1 \leqslant x \leqslant y \leqslant N$ fill up a matrix thus:

$$
\begin{vmatrix}
1^3 & 1^3 & 1^3+1^3 \\
1^3 & 2^3 & 1^3+2^3 \\
 & \vdots & \\
1^3 & N^3 & 1^3+N^3 \\
2^3 & 2^3 & 2^3+2^3 \\
 & \vdots & \\
2^3 & N^3 & 2^3+N^3 \\
3^3 & 3^3 & 3^3+3^3 \\
 & \vdots & \\
N^3 & N^3 & N^3+N^3
\end{vmatrix}
$$

(ii) Sort the rows into order using the third column as discriminator.

(iii) Scan the sorted third column looking for identical values. Each pair of such values represents a solution of the original equation. Having found all solutions for, say, $x, y, a, b \leqslant 50$, modify your program to eliminate all solutions which are simple multiples of other solutions (i.e. the highest common factor of $x, y, a, b > 1$).

7. Write a program to read in a number in 'Roman' style and print it out in 'Arabic' style. For example XLIV will yield 44 and so on.

8. The game of Nim is played by two players using three piles of matchsticks. Each player in turn takes a number of matches from one (and only one) pile. The player taking the last match or matches is the winner.

   The salient features of the game are:

   (i) If a *strategic* position is defined as one from which one player can force the other to lose, then strategic positions can occur in Nim and the rules for whether a position is strategic or not are as follows:

   Express the number of matches in each pile as a binary number and sum the digits without carry. If all these sums are even after the player has drawn, then the position is strategic for that player.

   (ii) If a position is strategic for A, then any move by B will make it non-strategic for B.

   (iii) If a position is non-strategic for A, then there is always a move which will make it strategic for B.

   Write a program to play the game from a starting position 5, 43, 46.

9. A Californian salesman's round involves 5 towns: Sacramento, Berkeley, Palo Alto, San Jose and Los Angeles. They are separated by the distances in miles given in the table below:

	S	B	PA	SJ	LA
Sacramento	0	85	117	163	489
Berkeley	85	0	30	50	385
Palo Alto	117	30	0	27	322
San Jose	163	50	27	0	302
Los Angeles	489	385	322	302	0

Calculate the shortest route which enables him to take in all the towns, starting and finishing in Sacramento.

11

Additional Language Features

Introduction

Many programmers will probably continue to write Fortran 90 programs in a style not too far removed from that permitted by FORTRAN 77. In the previous chapters, improvements made in Fortran 90—program layout, better "DO" and "CASE" structures, dynamic arrays, array passing in subroutines, intrinsic array functions, array sectioning and so on—which should be irresistible to any reasonable programmer, have been described.

Fortran 90 does, however, provide facilities for more radical changes in programming style and some of these features are described in the following sections.

Numerical computation

The precision with which arithmetic is done by computers has always been a thorny problem. For example, in the 1960s, disputes arose in the engineering literature about the relative merits of various linear algebraic equation solution algorithms the source of the disputes ultimately being traced to the different word lengths by which different computers represented numbers. In the previous chapters we have declared and used REAL, INTEGER and COMPLEX variables and have implicitly accepted that these will be represented internally in different computers by some "default" mechanism.

In Fortran 90 it is possible to specify the precision with which numbers will be represented (assuming of course that this is physically possible on a given processor). The only control which could be exercised in FORTRAN 77 was the DOUBLE PRECISION declaration for REALs which used two words (often 64 binary bits) to represent numbers instead of the default one word (often 32 binary bits). The means

in Fortran 90 for altering precision from the default are the intrinsic FUNCTIONs
SELECTED_INT_KIND and SELECTED_REAL_KIND for INTEGERs and REALs
respectively (see Appendix 3).

The KIND of a REAL or INTEGER number is an extension of the **FORTRAN 77**
DOUBLE PRECISION idea and is most useful for REALs. For example KIND = 2 is
equivalent to DOUBLE PRECISION for declaration of REALs. That is, the
declarations

```
DOUBLE PRECISION A,B,C
REAL (KIND = 2) :: A,B,C
```

are equivalent, the second being the preferred Fortran 90 style. Other KINDs of
numbers may be permissible depending on the processor.

In **FORTRAN 77**, once REAL constants or variables had been declared as
DOUBLE PRECISION, assignments to them had to use the letter D in place of the
letter E as an exponent, for example

$$+ 56.25D - 5$$

in the case of the very first example we saw in Chapter 3. In Fortran 90 using the
KIND = specifier, we might write:

```
INTEGER, PARAMETER :: DOUBLE = 2
REAL (KIND = DOUBLE) :: A
A = 56.25E-5_DOUBLE
```

When a new processor is encountered, the Fortran 90 programmer should first
make use of the selection intrinsic FUNCTIONs as follows.

The SELECTED_REAL_KIND intrinsic FUNCTION returns the KIND parameter
necessary for a given decimal precision. For example, on a processor with 64-bit
words, 15-figure decimal precision is likely to be achieved with KIND = 1, whereas
on a 32-bit processor, the same decimal precision would probably be achieved with
KIND = 2.

The statement

```
IKIND = SELECTED_REAL_KIND (P = 15)
```

would return values of 1 or 2 for IKIND in these two circumstances. If the processor
cannot achieve 15-decimal place accuracy, a negative number would be assigned to
IKIND. Having established that the desired precision is possible, one could declare

```
INTEGER, PARAMETER :: IKIND = SELECTED_REAL_KIND (P = 15)
```

and subsequently declare variables as

```
REAL (KIND = IKIND) :: A,B,C
```

This is a little long-winded, and on many processors it will be sufficient to
determine the "word length" for a given decimal precision and then to request that
wordlength for all REAL variables by means of a system ("job control") statement.

Some intrinsic FUNCTIONs are helpful for establishing precisions and tolerances in computations. Amongst these are

HUGE (X)	Largest representable number
TINY (X)	Smallest representable non-zero positive number
PRECISION (X)	INTEGER equal to the maximum decimal precision (the maximum number of significant figures)
RANGE (X)	INTEGER equal to the order of magnitude of the largest representable number or, if greater, of the inverse of the smallest.
EPSILON (X)	Positive real number almost negligible in comparison with unity (i.e. usable for as small a number as the processor can support).

On a typical 'PC' processor, the output from these five intrinsic procedures was

$$3.4028235E + 38$$
$$1.1754944E\text{-}38$$
$$6$$
$$37$$
$$1.1920929E\text{-}07$$

A less frequently used selection for INTEGERs is SELECTED_INT_KIND (LOW) which returns the lowest value of the KIND parameter for INTEGERs with LOW digits. If LOW-digit INTEGERs cannot be represented, -1 is returned.

Derived data TYPES

It may useful to make collections of data under a single heading which in Fortran 90 can be the name of a user-defined data TYPE. For example in the engineering application of finite element analysis (see Smith and Griffiths, 1988*) each "element" carries with it various attributes, such as

 (i) an identification number in some global scheme
 (ii) the dimensionality of the element—1,2 or 3
 (iii) the number of nodes the element has
 (iv) the numbers of its nodes in some global scheme
 (v) the spatial coordinates of the nodes in some global scheme
 (vi) the number of physical properties specified for the element
(vii) the values of these physical properties.

Attributes (i) (ii) and (iii) would consist of simple INTEGERs, say IELNO, IDIM and NOD. Attribute (iv) would be an INTEGER array, probably of rank 1 and of length equal to the number of nodes the element has. Attribute (v) would be a REAL array, probably of rank 2 whose second dimension reflects the physical dimensionality of the element—usually 2- or 3-dimensional. Attribute (vi) would be a simple INTEGER, say IPROP, and attribute (vii) would probably be a REAL array of rank 1, and length equal to the number of physical parameters required, Young's Modulus E and Poisson's Ratio V for an elastic solid for example.

* I. M. Smith and D. V. Griffiths, *Programming the Finite Element Method* 2nd ed, John Wiley and Sons. 1988

We could contain all of this data within a `TYPE` statement as follows:

```
TYPE ELEMENT
  INTEGER : : IELNO
  INTEGER : : IDIM
  INTEGER : : NOD
  INTEGER : : NODE_NUMBERS (32)
  ! No more than 32 nodes are allowed in an element
  ! Note that TYPEs cannot be ''dynamic'' ie (NOD) is not allowed
  REAL : : COORDS (32,3)
  INTEGER : : IPROP
  REAL : : PROPS (20)
  ! No more than 20 properties allowed
END TYPE ELEMENT
```

If there are then many (typically hundreds or thousands) of two-dimensional elastic elements in a problem we may declare for example an array:

```
TYPE (ELEMENT) : : TOTAL_ELEMENTS (100)
```

which would reserve space for more than 15000 pieces of information about the elements.

To select components of `TOTAL_ELEMENTS` we use the percent character; for example the element number of the 50th element in the `TOTAL_ELEMENTS` scheme would be found via

```
TOTAL_ELEMENTS (50) % IELNO
```

and the `INTEGER` returned could be used in assignment statements such as

```
IDIFF = TOTAL_ELEMENTS (100)% IELNO−TOTAL_ELEMENTS (1)% IELNO
```

Similarly, assignments of the form

```
TOTAL_ELEMENTS (1) = ELEMENT (1,2,8, NODNOS, &
                      POSITIONS, 2, 1.E6,.3)
```

can be made as an example of a "TYPE constructor". Components of derived data TYPEs can themselves be previously specified derived TYPEs and so we could cast the previous example in a different way by specifying

```
TYPE PROPERTY
  INTEGER : : IPROP
  REAL : : PROPS (20)
END TYPE PROPERTY
```

and then

```
TYPE ELEMENT2
  INTEGER : : IELNO
  INTEGER : : IDIM
  INTEGER : : NOD
  INTEGER : : NODE_NUMBERS (32)
  REAL : : COORDS (32,3)
  TYPE (PROPERTY) : : PROP
END TYPE ELEMENT2
```

The 100 elements of the new TYPE could be declared as

```
TYPE (ELEMENT2) :: TOTAL_ELEMENTS2 (100)
```

and the Young's Modulus of the first element retrieved as

```
TOTAL_ELEMENTS2 (1) % PROP % PROPS (1)
```

Derived data TYPEs cannot automatically be manipulated like the five intrinsic data TYPEs (added etc) but operators can be user-defined using MODULEs (see below).

MODULES

These are a new feature of Fortran 90 and provide a facility present in other languages such as C. They take the form of program units which are separate from the main program in the way that FUNCTIONs and SUBROUTINEs are. However a MODULE contains no executable statements and is merely a list or collection of declarations which is made permanently accessible to the program unit which invokes it. Because use of a MODULE involves essentially a declaration, it is invoked at the very beginning of a program before any other declarations via a USE statement. In a more complicated usage, a MODULE may contain subprograms (procedures), the purpose of this being to group together procedures which are somehow related and which require access to common data. Indeed one purpose of the MODULE idea is to replace the more cumbersome COMMON declaration in earlier FORTRANs which is considered obsolete.

The simplest form of a MODULE program unit might be typically

```
MODULE LOTS_OF_PARAMETERS
   INTEGER :: N,LALFA,LP,ITAPE
   REAL,ALLOCATABLE :: ALFA (:), BETA (:)
END MODULE LOTS_OF_PARAMETERS
```

so that a main program and subprograms needing these declarations would require only the statement

```
USE LOTS_OF_PARAMETERS
```

occurring before any executable statements to make the parameters available.

Parts of a MODULE can be selected in a USE statement, for example

```
USE LOTS_OF_PARAMETERS, ONLY :: LALFA, LP
```

makes only LALFA and LP available to that program unit.

Variables can be renamed, perhaps to contract or expand them to clarify their specific use, for example

```
USE LOTS_OF_PARAMETERS, ONLY :: LP ⇒ LAST_POINT
```

Here "⇒" is the pointer symbol more fully explained below.

MODULE procedures

In the most straightforward application of this concept, a MODULE may contain
a series of procedures, for example FUNCTIONs, which can be used by other
program units. Thus MODULE procedures are a way of implementing the idea of
"libraries" of subprograms in Fortran 90.

The construction is, typically:

```
MODULE SPECIAL
  USE Statements
  Declaration Statements
  CONTAINS
  FUNCTION BESSEL (X)
  REAL :: X
       :
  END FUNCTION BESSEL
  FUNCTION GAMMA (X)
  REAL :: X
       :
  END FUNCTION GAMMA
  FUNCTION HANKEL (X)
  REAL :: X
       :
  END FUNCTION HANKEL
END MODULE SPECIAL
```

The listed special functions would then be made available to any program unit
headed by the statement

```
USE SPECIAL
```

A more sophisticated deployment of MODULE procedures allows the creation of
new data types and their manipulation. For example, in Chapter 10, we wrote
a SUBROUTINE to multiply two matrices together to yield a third (the matrix
product). We noted that this is quite a different operation from Y = A*B which for
arrays Y, A and B satisfying conformability requirements would merely yield the
element-by-element products of the components of A and B. Furthermore, although
the result of our matrix multiply SUBROUTINE was a single quantity (the array Y),
we did not accomplish the operation by a FUNCTION.

We can attack this problem by creating a user derived data TYPE called, say,
MATRIX. Then we can construct a matrix multiply FUNCTION perhaps called
MATRIX_MULT which operates on TYPEs MATRIX.

Having written a whole collection of matrix manipulation FUNCTIONs and
SUBROUTINEs we may wish to collect them together in a matrix "library" (see
Chapter 12). A way of doing this in Fortran 90 is to create a MODULE, say
MATRIX_LIBRARY_MODULE.

This procedure "library" would then be available to any program containing the
statement

```
USE MATRIX_LIBRARY_MODULE
```

amongst its declarations. In practice, libraries such as this will tend to be "external" precompiled versions linked to the calling program after its compilation. An EXTERNAL statement and INTERFACE block (Chapter 12) should then be used.

Operator overloading

The standard Fortran 90 operators such as $+,-,*$ and so on are defined in the language as being applicable to intrinsically defined operands. Users can extend the applicability of the standard operators, for example to apply to derived data TYPEs, by a process known as "operator overloading". For example, the * operator applied to arrays of intrinsic TYPE produces the not very useful result that for arrays satisfying conformability conditions

```
A = B*C
```

yields an array A containing the element-by-element products of B and C. But in *matrix* arithmetic, what we might want is that the * operator would produce the matrix product of its operands.

Now suppose we have defined the TYPE called MATRIX in the previous section, written the FUNCTION called MAT_MULT which uses it, and put it in MATRIX_LIBRARY_MODULE. Within the MODULE we then need an "interface block" of the form

```
INTERFACE OPERATOR (*)
   FUNCTION MAT_MULT (MATRIX_A, MATRIX_B) RESULT &
                                   (MATRIX_Y)
      TYPE (MATRIX) :: MATRIX_Y
   END FUNCTION MAT_MULT
END INTERFACE
```

Then if A, B and C are of TYPE MATRIX, the operation

```
A = B*C
```

will return the derived matrix product to matrix A. To do something with the result, it may be desirable to convert matrix A back into an array of intrinsic TYPE or to write a MODULE procedure for dealing with products like A (printing them etc).

User-defined operators

If it seems confusing to redefine intrinsic operators, special operator names can be given in the interface block. For example, to replace "*" by "times" we could write

```
INTERFACE OPERATOR (.TIMES.)
     :
END INTERFACE
```

when the user-defined operator name is enclosed by two full stops. The call to operate on matrices would then take the form

```
A = B .TIMES. C
```

Generic assignment

The assignment operator " = " can also be overloaded using an interface block, as follows:

```
INTERFACE ASSIGNMENT ( = )
  Various SUBROUTINEs
END INTERFACE
```

Note that in this case the interface procedures must be SUBROUTINEs and not FUNCTIONs. They must have two arguments, the first having INTENT (OUT) or INTENT (INOUT) and the second INTENT (IN). User-defined assignment operators are also permitted.

The PRIVATE and PUBLIC attributes

So far we have assumed that the purpose of a MODULE is to make its contents universally accessible. These contents are said to have the attribute PUBLIC. If some of the contents are not to be accessible outside their own MODULE, this can be achieved by giving them the attribute PRIVATE.

Typical declaration statements might be

```
REAL, PUBLIC :: AUTOMATIC (100)
INTEGER, PRIVATE :: NO, NEVER
```

POINTERS

The distinction between the value of some quantity (REAL, array etc) and where to locate it in the machine (its "address") is fundamental to automatic computation. Whereas quite a lot of space may be necessary to hold values (for example of the elements of large arrays) a very small amount of space is necessary to describe where to find these values. The process of finding the address of a variable, or "pointing" to it has been present in computer languages from the very beginning but has been explicitly recognised in languages such as C and occurs in Fortran 90 in the form of POINTERs and TARGETs.

We have already used the concept in procedures (subprograms) in Chapter 10. When we wrote and used a SUBROUTINE such as MATRIX_MULT which had as arguments the arrays to be multiplied together, the resulting array and (possibly) the shapes of these, all of these arguments were "dummies" which pointed to the addresses of the actual arrays and shapes to be manipulated at run time. At no place in the SUBROUTINE was space allocated for any variables (which would of course be very wasteful for large arrays). This "indirect addressing" was present in ALGOL 60 in the form of the "call by name" and in ALGOL 68 in the form of "referencing", but POINTERs allied with structures and procedures are a much more powerful concept.

As a very simple illustration of the use of POINTERs in Fortran 90, we could declare a POINTER, say P, and its TARGETs, say T1, T2, T3, T4, T5 as follows:

```
REAL, POINTER :: P
REAL, TARGET :: T1,T2,T3,T4,T5
```

The TARGETs are ordinary variables with space reserved for them, but the POINTER has no reserved space. At run time, it can be associated with any TARGET (of the same TYPE) by the POINTER assignment, for example.

```
P ⇒ T1
```

Thereafter, P "pretends" to be T1 or is its "alias", and can be used as an alternative to T1 in expressions and assignments

```
T5 = T1
```
and
```
T5 = P
```

being equivalent. If P is changed, as in

```
P = 2.0*P
```

this has the effect of changing T1. In order to find out whether a POINTER has been lined up on a TARGET, the intrinsic FUNCTION ASSOCIATED can be used. For example, with the above declarations, ASSOCIATED (P) has the value .TRUE. if P is targeted on any variable, and the value .FALSE. if it is not. Further, ASSOCIATED (POINTER = P, TARGET = T1) is .TRUE. only if P is specifically targeted on T1.

After a POINTER has "locked on" to a TARGET, it can be disassociated by the statement NULLIFY, for example NULLIFY (P).

A POINTER can point to another POINTER without the latter having been declared as a TARGET. Thus if we have

```
REAL, POINTER :: P1,P2
```

and

```
P1 ⇒ T1
P2 ⇒ P1
```

the effect is for both P1 and P2 to point to T1.

POINTERs to arrays

To show how POINTERs are a much more powerful concept than simple "indirect addressing", we can point to sections of a two-dimensional array as follows:

```
REAL, POINTER :: VECTOR (:)
REAL, TARGET :: ARRAY (:,:)
VECTOR ⇒ ARRAY (I,:)
VECTOR ⇒ ARRAY (:,J)
```

which would first associate VECTOR with the Ith row of ARRAY and then with the Jth column of ARRAY. POINTERs can have memory space allocated to them or removed from them using the statements ALLOCATE (P1) and DEALLOCATE (P1). Note however that the intrinsic inquiry FUNCTION called ALLOCATED is for testing arrays and not POINTERs.

POINTERs to derived data TYPEs (structures)

Earlier in this chapter we saw that a useful collection of data in finite element analysis is under the heading of a derived data TYPE called ELEMENT:

```
TYPE ELEMENT
  INTEGER :: IELNO, IDIM, NOD
  INTEGER :: NODE_NUMBERS (NOD)
  REAL :: COORDS (NOD, IDIM)
  INTEGER :: IPROP
  REAL :: PROPS (IPROP)
END TYPE ELEMENT
```

Then, as long as they have been targeted using

```
TYPE (ELEMENT), TARGET :: TELEM1, TELEM2
```

we can point to TARGET elements TELEM1 and TELEM2 using

```
TYPE (ELEMENT), POINTER :: PELEM1, PELEM2
```

and

```
PELEM1 ⇒ TELEM1
PELEM2 ⇒ TELEM1
```

and so on. The POINTER names can be used in assignments to components of the derived data TYPE in the expected way, for example

```
YOUNGS_MODULUS = PELEM1 % PROPS (1)
```

The FUNCTIONs ASSOCIATED and NULLIFY and the statements ALLOCATE and DEALLOCATE apply to derived data TYPEs as well as to intrinsics.

Exercises

1. Several players are dealt one card each. The deal is input as follows:

```
S4 WILLIAMS
HQ JONES
DA BLOGGS
C0 SMITH
:
X0
```

Note how, to get it into one character, "ten" is typed as "0". Ace counts high. When comparing two cards, if the values are equal then Spades > Hearts > Diamonds > Clubs. The list is terminated by a card of the fictitious suit "X".

Write a program to go through the list and, at the end, print out the cards which scored highest and second highest, and the names of the corresponding players.

Hints: Observe that the order Spades, Hearts, Diamonds, Clubs is in inverse alphabetical order of their first letter. Intrinsic procedures may be useful (you may assume that ABS gives a sensible ordering for letters and digits, and remember that " > " etc may be used directly on CHARACTERs).

2. The ancient game of Hanoi is played with a board on which are three cylindrical pegs (1,2,3) and a number of rings (say 3) of varying diameters which fit on the pegs (like in Hooplah).

   At the start of the game, all rings are on one peg (say 2) such that no ring is on top of one smaller than itself. The object of the game is to move all pieces to another peg (say 1), subject to two constraints:

   (i) Only one piece may be moved at once
   (ii) At no time may any ring be on top of a smaller one.

   Hint: The solution is as follows, where, for example, 2 → 1 means move the ring on peg 2 to peg 1

   2 → 1
   2 → 3
   1 → 3
   2 → 1
   3 → 2
   3 → 1
   2 → 1

   See if the three-ring moves can be broken down into several two-ring moves and hence consider a recursive procedure solution.

3. Repeat Example 4 of Chapter 10, putting the procedures in a MODULE.

12
Libraries of External
SUBROUTINES

Introduction

A great advantage of using an established language like Fortran is the wealth of existing software upon which programmers can draw. Ideally, such software is in the form of "portable" subroutines which can be readily incorporated in user programs. In the previous two chapters, it was shown how the new features of Fortran 90, such as dynamic arrays, have made portability very much easier to achieve than it was in FORTRAN 77. The "upward compatibility" of Fortran means that FORTRAN 77 libraries can be used directly by Fortran 90 programs. For users of these libraries, this means that rather clumsy parameter lists are likely to remain for some time, but in due course conversion to the neater Fortran 90 styles will be increasingly achieved.

There are many subroutine libraries available in Fortran but in this chapter only three are described to give the flavour of their scope. The first two libraries consist of basic "Mathematical and Statistical" routines whilst the third is an example of an "applications-orientated" library, which helps with the solution of the most commonly occurring problems in numerical linear algebra.

Mathematical and statistical subroutines

Typical commercially available and supported libraries of this kind contain in excess of 500 subroutines. In the context of Fortran 90 it is of interest that some library routines, sometimes called "utility" routines are now part of the Fortran language and therefore available as intrinsics. Examples would be routines which return the largest and smallest representable positive numbers, which were shown in the previous chapter to be available as HUGE(X) and TINY(X) respectively.

The two libraries now described cover much of the same ground but with somewhat different emphasis. From an inspection of the lists given below it will be clear that some of the exercises given in this book could be solved using library routines, for example for the solution of sets of linear or non-linear equations, ordinary differential equations and so on.

The NAG FORTRAN library

The development of this widely used library has now reached the 15th revision—"Mark 15". Given below is a list of the "chapters" into which the subroutines are divided together with the major subdivisions in these chapters, if applicable.

Chapter	Topic	Major Subdivisions
A02	Complex Arithmetic	Square root of a complex number Modulus of a complex number Quotient of two complex numbers
C02	Zeros of Polynomials	All zeros of a polynomial, by Grant and Hitchin's Method
C05	Roots of One or More Transcendental Equations	Zero of a continuous function of one variable Solution of a system of non-linear equations (easy to-use routines) Solution of a system of non-linear equations (comprehensive routines) Check user's routine for calculating first derivatives
C06	Summation of Series	Discrete Fourier transform, by F.F.T. algorithm Complex conjugate of complex data values, Hermitian sequence Circular convolution of two real vectors of period 2^m Acceleration of convergence of a sequence, by epsilon algorithm Sum of a Chebyshev series
D01	Quadrature	One-dimensional integrals Two-dimensional integrals over a finite region Multi-dimensional integrals Weights and abscissae for Gaussian quadrature rules Korobov optimal coefficients for use in D01GCF
D02	Ordinary Differential Equations	Initial value problems for a system of O.D.E.s Boundary-value problems for a system of O.D.E.s Second-order Sturm-Liouville problems

Chapter	Topic	Major Subdivisions
D03	Partial Differential Equations	Elliptic problems Parabolic problems in one space variable, using method of lines Triangulation of a plane region
D04	Numerical Differentiation	Derivatives up to order 14 of a function of a single real variable
D05	Integral Equations	Linear non-singular Fredholm equation of second kind
E01	Interpolation	Interpolating functions Interpolated values
E02	Curve and Surface Fitting	Minimax curve fit by polynomials Least squares curve fit Least squares surface fit L_1-approximation L_∞-approximation Pade approximants Evaluation of fitted functions Differentiation and integration of fitted functions Sort two-dimensional data into panels for fitting or evaluating bicubic splines
E04	Minimising or Maximising a Function	Function of a Single Variable Minimum of a Function of One Variable Functions of Several Variables Unconstrained minimum (easy-to-use routines) Unconstrained minimum (comprehensive routines) Minimum subject to simple bounds on the variables (easy-to-use routines) Minimum subject to simple bounds on the variables (comprehensive routines) Minimum subject to general non-linear constraints Unconstrained minimum of a sum of squares (easy-to-use routines) Unconstrained minimum of a sum of squares (comprehensive routines) Service routines
F01	Matrix Operations, Including Inversion	Matrix inversion Pseudo-inverse and rank of a real $m \times n$ matrix $(m \geqslant n)$ Matrix factorisations Reduction of generalised real symmetric eigenproblems to standard form

Chapter	Topic	Major Subdivisions
F01 (continued)	Matrix Operations, Including Inversion	Balance a matrix by diagonal similarity transformations Reduction of matrices to condensed form by similarity transformations Backtransformation of eigenvectors from those of reduced forms Matrix and vector arithmetic
F02	Eigenvalues and Eigenvectors	Matrix eigenvalue problems (black box routines) Matrix eigenvalue problems (specialised routines) Singular value decomposition of a real $m \times n$ matrix Singular value decomposition of a real bidiagonal matrix
F03	Determinants	Determinant evaluation (black box routines) LU-factorisation and determinant LL^T-factorisation and determinant Determinant of a complex Hermitian positive-definite matrix, after factorisation by F01BNF
F04	Simultaneous Linear Equations	Solution of simultaneous linear equations (black box routines) Solution of simultaneous linear equations after factorising the matrix of coefficients Least-squares solution of m real equations in n unknowns
F05	Orthogonalisation	Schmidt orthogonalisation of N vectors of order M Approximate 2-norm of a vector
G01	Simple Calculations on Statistical Data	Simple descriptive statistics, one variable Simple descriptive statistics, two variables, from raw data Frequency table from raw data Two-way contingency table analysis Lineprinter scatterplot of two variables Lineprinter scatterplot of one variable against Normal scores Lineprinter histogram of one variable Statistical distribution functions Calculation of normal scores

Chapter	Topic	Major Subdivisions
G02	Correlation and Regression Analysis coefficients	Pearson product-moment correlation
		"Correlation-like" coefficients (calculated about zero)
		Kendall's and/or Spearman's non-parametric rank correlation coefficients
		Simple linear regression with constant term
		Simple linear regression without constant term
		Multiple linear regression
		Service routines for multiple linear regression
G04	Analysis of Variance	One-way analysis of variance, subgroups of unequal size
		Two-way analysis of variance, cross-classification, subgroups of equal size hierarchical classification, subgroups of unequal size
		Three-way analysis of variance, Latin square design
G05	Random Number Generators	Pseudo-random real numbers from continuous distributions
		Pseudo-random integer from uniform distribution
		Pseudo-random logical value
		Pseudo-random permutation of an integer vector
		Pseudo-random sample from an integer vector
		Pseudo-random integer from reference vector
		Set up reference vector for generating pseudo-random integers
		Set up reference vector from supplied cumulative distribution
		Set up reference vector for multivariate Normal distribution
		Pseudo-random multivariate Normal vector from reference vector
		Set up reference vector for univariate ARMA time-series model
		Generate next term from ARMA time-series using reference vector set up by G05EGF

Chapter	Topic	Major Subdivisions
G05 (continued)	Random Number Generators	Initialise random number generating routines Save state of random number generating routines Restore state of random number generating routines
G08	Nonparametric Statistics	Tests of location Tests of dispersion Tests of fit Tests of association and correlation
G13	Time Series Analysis	Univariate series Bivariate series
H	Operations Research	Linear programming problem Find feasible point or vertex which satisfies linear constraints Quadratic programming, by Beale's method Integer linear programming, by Gomory's method with Wilson's cuts Transportation problem
M01	Sorting	Sort a vector, by Singleton's implementation of Quicksort Sort a vector and provide an index to the original order Provide an index to the sorted order of a vector, leaving the vector unchanged Sort the rows of a matrix on keys in an index column Sort the columns of a matrix on keys in an index row
P01	Error Trapping	Return value of error indicator, or terminate program with an error message
S	Approximations of Special Functions	$Tan(X)$ $Arcsin(X)$ $Arccos(X)$ $Tanh(X)$ $Sinh(X)$ $Cosh(X)$ $Arctanh(X)$ $Arcsinh(X)$ $Arccosh(X)$ Exponential integral, $E_1(X)$ Sine integral, $Si(X)$ Cosine integral, $Ci(X)$

Chapter	Topic	Major Subdivisions
S (continued)	Approximations of Special Functions	Gamma function Log Gamma function Cumulative normal distribution function, P(X) Complement of cumulative normal distribution function, Q(X) Error function, erf(X) Complement of error function, erfc(X) Dawson's integral Bessel functions Airy functions Modified Bessel functions Fresnel integrals Elliptic integrals
X01	Mathematical Constants	π Euler's γ
X02	Machine Constants	Smallest positive ε such that $1.0 + \varepsilon > 1.0$ Smallest representable positive real number Largest representable positive real number Largest negative permissible argument for EXP Largest positive permissible argument for EXP Smallest representable positive real number whose reciprocal is also representable Largest permissible argument for SIN and COS Base of floating-point arithmetic Largest representable integer Largest positive integer power to which 2.0 can be raised without overflow Largest negative integer power to which 2.0 can be raised without underflow Maximum number of decimal digits that can be represented Estimate of active-set size (on machines with paged virtual store) Switch for taking precautions to avoid underflow
X03	Innerproducts	Real innerproduct added to initial value, using basic or additional precision Complex innerproduct added to initial value, using basic or additional precision
X04	Input/Output Utilities	Return or set unit number for error messages Return or set unit number for advisory messages

The IMSL FORTRAN libraries

This second example of mathematical and statistical subroutine provision is com-
plemented by a "Special Function" library similar to, but more extensive than,
Chapter 5 of the NAG library. IMSL arrange their subroutines using the following
subdivisions:

Mathematical Applications
    Differential Equations; Quadrature; Differentiation
    Eigensystem Analysis
    Interpolation; Approximation; Smoothing
    Linear Algebraic Equations
    Linear Programming
    Nonlinear Equations
    Optimisation
    Transforms
    Vector/Matrix Arithmetic
Statistical Applications
    Analysis of Variance
    Basic Statistics
    Categorised Data Analysis
    Nonparametric Statistics
    Observation Structure; Multivariate Statistics
    Regression Analysis
    Sampling
    Time Series; Forecasting; Econometrics
General Applications
    Generation and Testing of Random Numbers
    Mathematical and Statistical Special Functions
    Utility Functions

In the case of the Special Functions Library, the subdivisions are as follows:

    Bessel Functions
    Bessel Functions of Fractional Order
    Elementary Functions
    Error Function and Related Functions
    Exponential Integrals and Related Functions
    Fundamental Functions
    Gamma Functions and Related Functions
    Miscellaneous Functions
    Trigonometric and Hyperbolic Functions

Despite the usefulness of these libraries it should be emphasised that if applicable,
Fortran 90 intrinsics, for example for array manipulation, should always be
preferred since they are likely to be machine-optimal.

LAPACK—an applications orientated library

It will be clear from the preceding chapters and examples that linear algebra forms a major applications area for engineering and scientific programs. In the 1970s, two libraries were developed, called LINPACK and EISPACK, to deal with linear algebraic and eigensystems respectively. These have since been merged into a single library, called LAPACK, by the Society of Industrial and Applied Mathematics (SIAM) in the USA and NAG Ltd (the vendors of the NAG mathematical subroutine library previously described) in the UK.

LAPACK can solve systems of linear equations, linear least squares problems, eigenvalue problems and singular value problems. It consists of "driver" routines, "computational" routines and "auxiliary" routines.

"Driver" routines solve a complete problem, for example a system of linear equations or the eigenvalues of a real, symmetric matrix. A wide variety of matrix types and storage strategies is provided. For example, in the case of linear equation system solution, the following matrix types and storage schemes are allowed:

1. General
2. General band
3. General tridiagonal
4. Symmetric/Hermitian positive definite
5. Symmetric/Hermitian positive definite (packed storage)
6. Symmetric/Hermitian positive definite band
7. Symmetric/Hermitian positive definite tridiagonal
8. Symmetric/Hermitian indefinite
9. Complex Symmetric
10. Symmetric/Hermitian indefinite (packed storage)
11. Complex Symmetric (packed storage)

"Computational" routines perform a distinct computational task, like the LU factorisation described in Chapter 13, again allowing for several different matrix types and storage strategies.

"Auxiliary" routines perform subtasks at a lower level, such as returning various norms of matrices, rotations and so on.

The essential point is that so much FORTRAN code already exists, in many applications areas, that it makes sense to utilise it wherever possible.

Graphics libraries in FORTRAN

The field of computer graphics provides an object lesson in the need for standardisation. Although graphics libraries can be written in FORTRAN (for example the NAG Graphics Library, currently at Mark 3) an interface must still be provided to the particular plotting package available on the user's hardware. As an example, the NAG Graphics Library provides standard interfaces to the following plotting packages:

Adobe PostScript
Tektronix PLOT 10 IGL (Level 6)
DEC ReGIS
Hewlett Packard HPGL
Lineprinter
GKS (Version 7.4)
Cal Comp HCBS
GINO—F (Version 2.7)
GHOST80

and there are many others. The main point about writing graphics software in FORTRAN is that at least portability can be achieved at the level of the calling program.

To preserve maximum portability, a good strategy is to call only the most basic level of graphics subroutines from the (Fortran 90) program. These are often called "Plotting Primitives" and can be found in graphics libraries. For example the NAG Graphics Library contains the following primitives:

- move the pen to the position (x, y) in user coordinates
- increment the pen position by an amount $(\delta x, \delta y)$ in user coordinates
- draw a line from the current pen position to the position (x, y) in user coordinates
- draw a line from the current pen position, advancing by an increint (x, y) in user coordinates
- draw a marker
- draw a character string
- set the marker size
- set the width and height of characters
- set the character spacing
- set the current pen
- perform area fill of a "closed" polygon
- set the current brush
- set the colour index
- set the current line style
- set the current text font.

Indeed, some Fortran 90 systems contain a selection of these primitives as intrinsic FUNCTIONs and so a separate library is not necessary.

Incorporating subprogram libraries in Fortran 90 programs

In Chapter 10 we saw a simple way of "containing" subprograms within a Fortran 90 main program. The subprograms, written in Fortran 90, followed the keyword CONTAINS after the last main program statement before END. In this style it was perfectly possible for subprograms to contain other subprograms.

Following the introduction of MODULEs in Chapter 11, another way of incorporating subroutines in a main program suggests itself. In this method, a MODULE containing the subprograms, written in Fortran 90, precedes the main program. The statement USE, with the name of the subprogram MODULE must then occur amongst the declarations in the main program. A typical program structure is illustrated below:

```
MODULE PROCEDURES
CONTAINS
SUBROUTINE STATISTICS ( )
    —

    —

    —
END SUBROUTINE STATISTICS
REAL FUNCTION CORRELATION ( )
    —

    —

    —
END FUNCTION CORRELATION
END MODULE PROCEDURES
```

```
PROGRAM STATISTICS_PROCEDURES
USE PROCEDURES
    —

    —

    —
CALL STATISTICS ( )
    —

    —

    —
END PROGRAM STATISTICS_PROCEDURES
```

A third way of incorporating subprograms, deprecated by some because there is a chance of loss of portability, is to use an INCLUDE line before the main program. This method takes the form:

```
INCLUDE 'library.mod'
```

```
PROGRAM MAIN
USE LIBRARY
    —

    —

    —

    —
END PROGRAM MAIN
```

Here a file 'library.mod' contains the Fortran 90 subprograms which constitute the library. The name LIBRARY in the USE statement in the main program is the name in the first line of the 'library.mod' file. This can lack portability because there may be system-dependent restrictions on the style of the filename such as 'library.mod'.

In all of the above examples, we have assumed that the subprograms are written in Fortran 90. If there are lots of subprograms in a library, it is inefficient to recompile them each time a program is run, and what we want to be able to do is to gain access to a file containing precompiled versions of the library routines.

This can be done most safely by keeping all the library SUBROUTINEs (and FUNCTIONs) in a Fortran 90 file and precompiling them. To load and link this file with the calling program we again whole make use of a MODULE. In this context, the MODULE (let us call it INTERFACES) merely contains the "headers" of the SUBROUTINEs in the library as follows:

```
MODULE INTERFACES
   INTERFACE
      SUBROUTINE ONE (----)
      END SUBROUTINE ONE
          ⋮
   END INTERFACE
END MODULE INTERFACES
```

Listed in the MODULE are merely the names of each SUBROUTINE together with its parameter list and the TYPES of these parameters. This MODULE is then compiled in Fortran 90 and any calling program gains access to the SUBROUTINEs by the inclusion of the statement USE INTERFACES before any executable statements. It should be possible for systems to load and link only the selected SUBROUTINEs called by that particular main program. This method is quite safe, since the compiler can check the numbers and TYPEs of all the arguments.

13
Case Studies

Introduction

Now that the features of Fortran 90 have been explained, some longer examples are given in this chapter as "case studies". Solutions are as usual given in Appendix 5.

Case study 1: Newton–Raphson

If $y = f(x)$, to find y for a given x is usually simple. However to find x for a given y, we must often resort to "iteration" in which a guess is made for x. Using a method called "Newton–Raphson" iteration we recast the original equation as $f(x) - y = 0$ or $g(x) = 0$ and then for a starting guess x_0 we seek a better approximation by using the formula

$$x_1 = x_0 - \frac{g(x_0)}{g'(x_0)}$$

where $g' \equiv d/dx$, the first derivative. Having found x_1, we use it in place of x_0 to get an even better approximation, x_2, and so on.

Write FUNCTIONs to evaluate g and g' for the function given in Chapter 10, exercise 5, and hence write a program to calculate the strain for a stress of 1.5 Pa.

Hints: (a) In any iterative process, a "counter" should be used to stop the calculation if no solution is found within a reasonable number of trials, say 30.
   (b) It will be (almost) impossible to find the solution "exactly". The solution will be "good enough" if x_n and x_{n-1} differ by less than 10^{-6}.

Case study 2: bisection

The previous study assumed that g' could be found analytically. The "bisection" iteration can be carried out even if this is not so. Two initial approximations, x_1 and

x_2 are made and $g(x)$ determined at both points. If the two approximations bracket the desired solution, as below

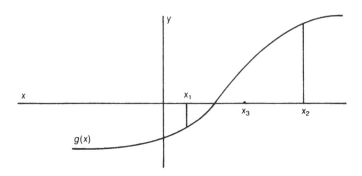

their g values will have opposite signs. The bisector of x_1 and x_2, or mid-point, is then assumed to be a better approximation to $g(x) = 0$. It then replaces whichever of x_1 and x_2 has the same sign and the iteration continues.

Write a program to solve the previous exercise by means of the bisection method.

Case study 3: solution of linear equations

The "Gaussian Elimination" process for solving sets of linear simultaneous equations

$$Ax = b$$

where A is an $N \times N$ matrix and x and b are N-long vectors is usually implemented in the form of a "factorisation"

$$A = LU$$

where L and U are lower and upper triangular matrices respectively (see for example Griffiths and Smith, 1991, p16*). Having found L and U the solution is completed by a "forward substitution"

$$Ly = b$$

and a "backward substitution"

$$Ux = y$$

Write a SUBROUTINE to factorise A and do the forward and backward substitutions for a given b returning result x.

*Griffiths, D.V. and Smith, I.M. *Numerical Methods for Engineers*, Blackwell, 1991

Case study 4: deflections of a truss

The "pin-jointed" engineering structure shown below is composed of linear elastic lines connected at "nodes" 1 to 10 as shown. Both the load and the displacement at any node may have an x and a y component. These components are given the suffixes shown as $(\,,)$ appended to each node (except 1 and 9 which cannot move in either direction).

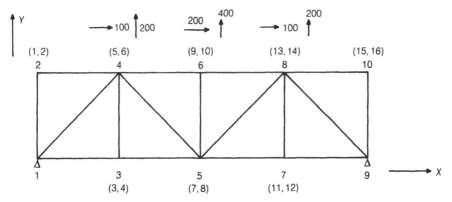

The load on the structure can be written as a vector b and the deflection as a vector x and to find x for a given b we have to solve the linear system of equations

$$Ax = b$$

Use the SUBROUTINE from Exercise 3 to solve this problem for loadings at nodes 4, 6 and 8 only, with the following values:

$$b^5 = 100\,\text{N}, \qquad b^9 = 200\,\text{N}, \qquad b^{13} = 100\,\text{N}$$
$$b^6 = 200\,\text{N}, \qquad b^{10} = 400\,\text{N}, \qquad b^{14} = 200\,\text{N}$$

The "stiffness" matrix A for the structure is shown below:

$$
\begin{bmatrix}
100 & 0 & 0 & 0 & -100 & 0 \\
0 & 100 & 0 & 0 & 0 & 0 & & & & & & 0 \\
0 & 0 & 200 & 0 & 0 & 0 & -100 \\
0 & 0 & 0 & 100 & 0 & -100 & 0 \\
-100 & 0 & 0 & 0 & 264 & 0 & -32 & 24 & -100 \\
0 & 0 & 0 & -100 & 0 & 136 & 24 & -18 & 0 \\
 & & -100 & 0 & -32 & 24 & 264 & 0 & 0 & 0 & -100 & 0 & -32 & -24 \\
 & & & & 24 & -18 & 0 & 136 & 0 & -100 & 0 & 0 & -24 & -18 \\
 & & & & -100 & 0 & 0 & 0 & 200 & 0 & 0 & 0 & -100 & 0 \\
 & & & & & & 0 & -100 & 0 & 100 & 0 & 0 & 0 & 0 \\
 & & & & & & -100 & 0 & 0 & 0 & 200 & 0 & 0 & 0 \\
 & & 0 & & & & 0 & 0 & 0 & 0 & 0 & 100 & 0 & -100 \\
 & & & & & & -32 & -24 & -100 & 0 & 0 & 0 & 264 & 0 & -100 & 0 \\
 & & & & & & -24 & -18 & 0 & 0 & 0 & -100 & 0 & 136 & 0 & 0 \\
 & & & & & & & & & & & & -100 & 0 & 100 & 0 \\
 & & & & & & & & & & & & 0 & 0 & 0 & 100
\end{bmatrix}
$$

Is there scope for a more efficient solution of this problem?

Case study 5: conjugate gradient method

A second method for solving linear simultaneous equations (see for example Griffiths and Smith, 1991, p54*) is called the method of "conjugate gradients". It is an iterative procedure and arrives at a solution of

$$Ax = b$$

by the following steps:

$$p^0 = r^0 = b - Ax^0$$

where x^0 is a starting guess at the solution and p and r are vectors. Then for the kth iteration:

$$u^k = Ap^k$$

$$\alpha_k = \frac{(r^k)^\mathsf{T} r^k}{(p^k)^\mathsf{T} u^k}$$

$$x^{k+1} = x^k + \alpha_k p^k$$

$$r^{k+1} = r^k - \alpha_k u^k$$

$$\beta_k = \frac{(r^{k+1})^\mathsf{T} r^{k+1}}{(r^k)^\mathsf{T} r^k}$$

$$p^{k+1} = r^{k+1} + \beta_k p^k$$

where α_k and β_k are scalar numbers and u is a vector. In perfect arithmetic, the solution is reached in N iterations of this process where N is the number of equations being solved. Write a program to employ the above algorithm to solve a set of equations and test it on the problem described in case study 4. Your program should stop when the solution is "accurate enough" i.e. the difference between successive x is "small enough" or a maximum number of iterations has been reached.

Is there scope for more efficient iterative solutions?

Case study 6: equations with complex coefficients

Use the method of case study 3 to solve the COMPLEX system $Ax = b$ where

$$A = \begin{pmatrix} (1.0,\ 0.0) & (1.0,\ 0.0) & (0.0,\ 1.0) \\ (1.0,\ 0.0) & (0.0,\ 1.0) & (1.0,\ 0.0) \\ (0.0,\ 1.0) & (1.0,\ 0.0) & (1.0,\ 0.0) \end{pmatrix}$$

$$b = ((5.0,\ 0.0)\quad (0.0,\ -2.0)\quad (1.0,\ 0.0))^\mathsf{T}$$

*Griffiths, D.V. and Smith, I.M. *Numerical Methods for Engineers*, Blackwell, 1991.

Case study 7: eigenvalues

Consider vibration of three railway wagons as shown. For a force f_n applied to wagon n, we have static displacements given by

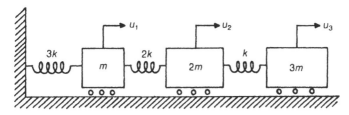

Figure. Vibrational system with three degrees of freedom

$$u_1 = \frac{1}{3k}f_1 + \frac{1}{3k}f_2 + \frac{1}{3k}f_3$$

$$u_2 = \frac{1}{3k}f_1 + \frac{5}{6k}f_2 + \frac{5}{6k}f_3$$

$$u_3 = \frac{1}{3k}f_1 + \frac{5}{6k}f_2 + \frac{11}{6k}f_3$$

or

$$\begin{pmatrix} u_1 \\ u_2 \\ u_3 \end{pmatrix} = \begin{pmatrix} \frac{1}{3k} & \frac{1}{3k} & \frac{1}{3k} \\ \frac{1}{3k} & \frac{5}{6k} & \frac{5}{6k} \\ \frac{1}{3k} & \frac{5}{6k} & \frac{11}{6k} \end{pmatrix} \begin{pmatrix} f_1 \\ f_2 \\ f_3 \end{pmatrix}$$

or $u = Mf$ where M is the "influence matrix" and its elements are "influence coefficients".

If no external forces are applied, we have $f_n = -m_n \ddot{u}$ (where $m_1 = m, m_2 = 2m, m_3 = 3m$).

For a natural vibration,

$$u_n = x_n \sin(\omega t + \phi)$$
$$\therefore \ddot{u}_n = -\omega^2 x_n \sin(\omega t + \phi)$$
$$\therefore f_n = m_n \omega^2 x_n \sin(\omega t + \phi)$$

$$\therefore \begin{pmatrix} x_1 \\ x_2 \\ x_3 \end{pmatrix} = \begin{pmatrix} \frac{1}{3k} & \frac{1}{3k} & \frac{1}{3k} \\ \frac{1}{3k} & \frac{5}{6k} & \frac{5}{6k} \\ \frac{1}{3k} & \frac{5}{6k} & \frac{11}{6k} \end{pmatrix} \begin{pmatrix} m\omega^2 x_1 \\ 2m\omega^2 x_2 \\ 3m\omega^2 x_3 \end{pmatrix} = \frac{m\omega^2}{k} \begin{pmatrix} \frac{1}{3} & \frac{2}{3} & 1 \\ \frac{1}{3} & \frac{5}{3} & \frac{5}{2} \\ \frac{1}{3} & \frac{5}{3} & \frac{11}{2} \end{pmatrix} \begin{pmatrix} x_1 \\ x_2 \\ x_3 \end{pmatrix}$$

or $X = \lambda GX$ where $\lambda = m\omega^2/k$ and $1/\lambda$ is an eigenvalue of G (or λ is an eigenvalue of G^{-1}).

We wish to find the lowest natural frequency of the system. In this mode, we expect wagon 3 to have the greatest displacement. Therefore, let us try

$$X = \begin{pmatrix} 0 \\ 0 \\ 1 \end{pmatrix}$$

If we now compute $X^* = GX$, and scale X^* so that its third element is 1, we may check whether X^*, as scaled, is the same as X(or, in practice, sufficiently close). If so, X is an eigenvector and the scaling factor is the eigenvalue $1/\lambda$. Otherwise, we try again using X^* in place of X. It can be shown that the process converges on the eigenvalue $1/\lambda$ of largest absolute value, corresponding to the lowest value of ω, as required.

Write a program to find the lowest natural frequency $= (\omega/2\pi)$ of a system of n railway wagons, given their masses $(m_1, m_2, \ldots, m_n)$, the stiffness of their springs $(k_1, k_2, \ldots, k_n)$, and an initial approximation to the eigenvector. Test your procedure on the three wagon problem above.

Case study 8: Runge-Kutta

A popular method for solving systems of (non-linear) ordinary differential equations is the "Runge-Kutta" method of the fourth order (see for example Griffiths and Smith, 1991, p 226*). In general, such a system will be of the form

$$\frac{dy_i}{dx} = f_i(x, y_0, y_1, \ldots, y_{n-1}) \qquad i = 0, 1, 2, \ldots, n$$

with n initial conditions $y_i(x_0) = A_i$, $i = 0, 1, 2, \ldots, n$.

For example the system of two equations:

$$\frac{dy}{dx} = f(x, y, z) \qquad y(x_0) = y_0$$

$$\frac{dz}{dx} = g(x, y, z) \qquad z(x_0) = z_0$$

may be solved by advancing the solution of y and z to $x_1 = x_0 + h$ by the formulae

$$y(x_1) = y(x_0) + K$$
$$z(x_1) = z(x_0) + L$$

---

* Griffiths, D.V. and Smith, I.M. *Numerical Methods for Engineers*, Blackwell, 1991

where $K = (K_0 + 2K_1 + 2K_2 + K_3)/6$ and $L = (L_0 + 2L_1 + 2L_2 + L_3)/6$ and

$$K_0 = hf(x_0, y_0, z_0)$$
$$L_0 = hg(x_0, y_0, z_0)$$
$$K_1 = hf(x_0 + \tfrac{1}{2}h, y_0 + \tfrac{1}{2}K_0, z_0 + \tfrac{1}{2}L_0)$$
$$L_1 = hg(x_0 + \tfrac{1}{2}h, y_0 + \tfrac{1}{2}K_0, z_0 + \tfrac{1}{2}L_0)$$
$$K_2 = hf(x_0 + \tfrac{1}{2}h, y_0 + \tfrac{1}{2}K_1, z_0 + \tfrac{1}{2}L_1)$$
$$L_2 = hg(x_0 + \tfrac{1}{2}h, y_0 + \tfrac{1}{2}K_1, z_0 + \tfrac{1}{2}L_1)$$
$$K_3 = hf(x_0 + h, y_0 + K_2, z_0 + L_2)$$
$$L_3 = hg(x_0 + h, y_0 + K_0, z_0 + L_2)$$

Write a program to solve the pair of equations

$$\frac{dy}{dx} = 3xz + 4, \qquad y(0) = 4$$

$$\frac{dz}{dx} = xy - z - e^x, \qquad z(0) = 1$$

by this method at $x = 0.5$, using steps of 0.1.

Case study 9: water levels in a turbine

A "surge tank" is a device for damping out the effects of suddenly starting or stopping a turbine in the arrangement shown below:

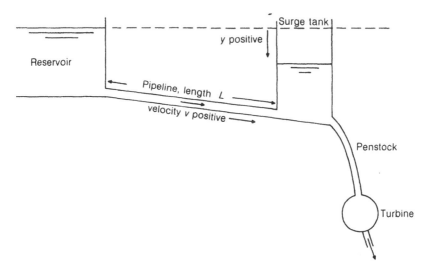

With normal flow through the turbine, the level in the surge tank is below the reservoir level by the amount of friction head loss h_f in the pipeline, so that initially $y = h_f = 4fLv^2/2gd$. If the turbine is suddenly shut down the flow in the penstock is stopped and the momentum of the water in the pipeline carries it into the surge tank causing the water level to rise. A pressure gradient builds up against the pipeline flow and decelerates it. The motions of the water in the pipeline and the water level in the surge tank are described by the equations:

$$A\frac{dy}{dt} = -va + X_t$$

$$\frac{L}{g}\frac{dv}{dt} = y - h_f$$

where X_t is the rate of flow through the turbine (zero after shut-down). Typically the level in the surge tank will oscillate as shown below:

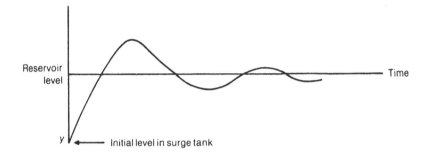

Given that the reservoir level remains constant, and that

$$L = 1200\,m$$
$$d = 800\,mm$$
$$D = 2.5\,m$$
$$f = 0.005$$
$$X_t\ (\text{before shut-down}) = 1\,m^3/\text{sec}$$

use the Runge-Kutta method of case study 8 to calculate what will happen

(a) when the turbine is shut down after steady running
(b) when the turbine is started after long quiescence.

Hence design a suitable height of surge tank.

Case study 10: date of Easter

Easter Day in any given year is the Sunday following the first Full Moon on or after March 21st. A Full Moon is deemed to occur 13 days after the preceding New Moon. The date of the last New Moon in January may be calculated from the "Epact" for the year, and from that the new moons in February, March and April may be obtained. The exact method is given in the Structure Chart below. All dates are held as the number of days from the beginning of the year.

Write a program to calculate and print the date of Easter in the year 7086.

LEAP = 1 if YEAR is a leap year, and 0 otherwise.
    (YEAR is a leap year if it is divisible by 4 but not by 100, or if it is divisible by 400)

| compute the DOMINICAL NUMBER for YEAR |

| compute the GOLDEN NUMBER for YEAR |

| compute the EPACT for YEAR |

CLAVIAN
CORRECTION =

GOLDEN NUMBER	
> 11	< = 11
26	25

JANUARY MOON = 31 − EPACT
  [2nd Jan < JANUARY MOON < = 31st Jan]

MARCH MOON = JANUARY MOON + 59 + LEAP
  2nd Mar < = MARCH MOON < = 31st Mar]

APRIL MOON = MARCH MOON +

EPACT	
> = CLAVIAN CORRECTION	< CLAVIAN CORRECTION
30	29

[1st Apr < = APRIL MOON < = 29th Apr]
PASCHAL MOON =

does the Full Moon following MARCH MOON fall before March 21st?	
Y	N
APRIL MOON + 13	MARCH MOON + 13

PASCHAL DAY = (PASCHAL MOON-DOMINICAL NUMBER).
MOD 7 + 1
  [1 = Sunday,..., 7 = Saturday]

EASTER = PASCHAL MOON + 8 − PASCHAL DAY

Print EASTER as MONTH (March or April) followed by DAY (followed by ST, ND, RD, or TH if you can manage it)

COMPUTE THE DOMINICAL NUMBER

> To find the Dominical or Sunday Number for any given Year, add to the year its fourth part, omitting fractions; next subtract the number of hundreds contained in that given year; next add the fourth part, omitting fractions, of that number of hundreds and also subtract 1 if the year be not Bissextile, and 2 otherwise: Divide by 7, and subtract the remainder from 7 to obtain the dominical number.
>
> [The dominical number gives the date of the first Sunday in January. Note that "Bissextile" means exactly the same as "Leap"]

TO COMPUTE THE GOLDEN NUMBER

> To find the Golden Number, or Prime, add one to the Year and then divide by 19; the remainder, if any, is the Golden Number; but if nothing remaineth, then 19 is the Golden Number.

TO COMPUTE THE EPACT

> Compute the Lilian correction
>
> To find the Epact for any given Year, multiply the Golden Number diminished by 1 by 11 and subtract the Lilian correction: the Epact is the result taken modulo 30.
>
> [The Epact is the age of the moon (i.e. the number of days since the New Moon) on January 1st.]

TO COMPUTE THE LILIAN CORRECTION

> Let 'century' be the number of hundreds contained in the given year.
>
> The Lilian correction is given by
> (century-century/4-(century-(century-17)/25)/3-8) taken modulo 30. [Ignore all remainders with the = operator]

Case study 11: calendar

Zeller's method for calculating the day of the week corresponding to a given date $d/m/y$ in the "Gregorian" calendar is as follows:

(i) If $m = 1$ or 2, add 12 to m and subtract 1 from y.
(ii) Let r be the remainder on dividing by 7 the expression

$$d + 2m + 2 + [(3m + 3)/5] + y + [y/4] - [y/100] + [y/400]$$

where [ ] indicates remainders are ignored after division operations
(iii) If $r = 0$ the day is a Saturday, if $r = 1$ the day is a Sunday and so on until if $r = 6$ the day is a Friday.

Using this method as the basis, write a program to output a complete calendar for a given year.

You may wish to include a check that the generated $d/m/y$ is a valid Gregorian date.

Case study 12: reverse Polish notation

In the usual notation of algebra, operators are placed between the operands on which they operate. For example

$$a + b \qquad x + y*z \qquad x*y + z$$

In this notation, precedence of operators is established as follows:
$$**$$
$$*/$$
$$+ -$$

so that the second example means 'the sum of x and the product of y and z'. To override this precedence, brackets must be used. Thus

$$(x + y)*z$$

means 'the product of the sum of x and y with z'.

In Reverse Polish notation, the operator comes after the operands on which it operates. Thus in Reverse Polish, the above examples appear as:

$$ab +$$
$$xyz* +$$
$$xy*z +$$
$$xy + z*$$

respectively. There is no need for brackets (parentheses).

Write a program which will read in an expression in 'normal' notation and print out the Reverse Polish equivalent. Single letter operands will be sufficient.

Case study 13: Ruritanian stamps

In Ruritania, they have two simple postal regulations:

(i) Postage is at a flat rate of one cent per gram.
(ii) Only one, two or three stamps may appear on a letter or parcel.

A new set of n stamps is to be printed. The problem is to write a program to choose the n stamps in such a way that the range of continuous postage rates from one cent upwards is maximised.

For example, if $n = 4$, then a choice of 1 2 4 8 would give a maximum of 14 $(8 + 4 + 2)$, because although 16 is possible $(8 + 8)$, 15 is not. The solution in this case is 1 4 7 8, which gives a maximum of 24.

First thoughts when presented with a problem like this are towards deriving an analytical solution, since if this is possible the computation to be done is practically nil. No such solution has so far been found.

Thus we shall adopt a more direct approach: test all possible sets of stamps to find the best.

The program can be described by means of a Structure Chart (see below). 'Initialise' does all the usual tasks such as initialising 'best set' and 'best max' and producing the first set to be tried.

'Find maximum value of set' does what its title suggests: it finds the maximum range of continuous rates called 'max value of set' for any set presented to it.

If it is greater than the current 'best max', then 'best max' and 'best set' are updated.

'Generate next set if any' again is fairly self descriptive. It generates the next set to be tried. If there is one, we jump back to test it. If not, then 'Finalise' prints out the solution.

Clearly, all the problems are found in these two parts: 'Find max value of set' and 'Generate next set if any'. And of course they can both be written as procedures.

Consider first a procedure to 'Find max value of set'.

We shall store the set of stamps in an INTEGER array SET (0 : N) such that:

```
SET (0)    smallest
SET (1)    next smallest
  :
SET (N)    largest
```

The procedure will produce from these the maximum value. Its heading could be:

```
INTEGER FUNCTION GET_MAX_VAL (SET,N) RESULT (MAX_VAL)
```

Two methods of finding this maximum value suggest themselves.

First, see whether we can create a one cent postage using no more than three stamps. Then try for two cents. Then for three cents. And so on. When we discover one that we cannot create with up to three stamps, then the maximum is one less than this. We note that this solution is rather slow, since the process is an indirect one.

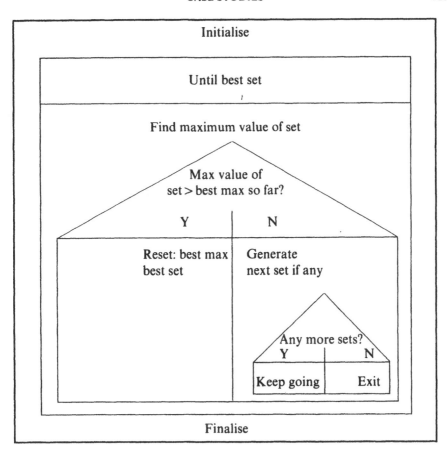

The second method is a direct one: we create all possible postages that can be created by one, two or three stamps on a letter. We then scan these to find the maximum.

Suppose, then, that we have a LOGICAL array POSTAGE and we associate the ith element of it with a postage of i cents. Then we can find the maximum as shown below.

To calculate the possible postages involves calculating all postages with only one stamp, all postages with two stamps, and all postages with three stamps, see below.

The first involves a statement with a single DO loop; the second involves a statement with two nested DO loops, one controlling the first stamp, the other the second. The third involves a statement with three nested DO loops.

By means of a simple dodge we can coalesce these three into one. Suppose each set of stamps is augmented by a further (notional) stamp of 0 cents' denomination. Call it SET (0). Then a postage with, say, one stamp can be considered as a postage of three stamps, two of which are of 0 cents.

Set all POSTAGE to .FALSE.

Calculate all possible postages
and set POSTAGE to .TRUE.

Until POSTAGE (I) is .FALSE.

Scan POSTAGE

Maximum value of SET $= i - 1$

Calculate 1-stamp postage

Calculate 2-stamp postage

Calculate 3-stamp postage

Now a simple analysis shows that the statement:

```
POSTAGE (SET(I) + SET(J) + SET(K)) = .TRUE.
```

is obeyed, for $n = 4$, a total of $5 * 5 * 5 = 125$ times, and that this dominates the calculation. Is this an irreducible minimum? A few moments' thought shows that if the set were, as usual, 1 2 4 8, then we test postages corresponding, for example, to the combinations of stamps:

$$
\begin{array}{ccc\quad ccc}
1 & 2 & 4 & 1 & 4 & 2 \\
2 & 1 & 4 & 2 & 4 & 1 \\
4 & 1 & 2 & 4 & 2 & 1
\end{array}
$$

so there is potentially a saving of something like six.

Now in evaluating a postage the procedure chooses each stamp of the three independently of the others. Suppose we choose the stamps in a more orderly fashion by choosing the smallest first, the next smaller (or equal smallest) second, and the largest third. Then of the six combinations above we consider only the first, namely 1 2 4.

We can do this by simply altering this structure from:

```
for I from 0 to N do
for J from 0 to N do
for K from 0 to N do
POSTAGE (SET(I) + SET(J) + SET(K)) = .TRUE.
```

to:

```
for I from 0 to N do
for J from I to N do
for K from J to N do
POSTAGE (SET(I) + SET(J) + SET(K)) = .TRUE.
```

This reduces the number of times this statement:

```
POSTAGE (SET(I) + SET(J) + SET(K)) = .TRUE.
```

to 35 from 125, a saving of almost a factor of 4.

So, given a set of stamps, we have a procedure for calculating the maximum value of the set. Now we must consider the problem of generating a set of stamps. Clearly, we cannot investigate every set, since there are an infinite number. We can, however, limit the number of sets to be considered.

SET (1) = 1 otherwise one cent is impossible. Lower limit of SET (2) = 2 and upper limit = 4 otherwise four cents is impossible. Beyond this the limit of the value of a stamp is dependent on the actual value of the previous stamp.

```
SET(1) = 1
SET(2) = 2 → 4
    ⋮
SET(N) = SET(N−1) + 1 → 3 * SET(N−1) + 1
```

Initialise sets up initial values in SET which are the upper limits above. If N = 4

```
SET(1) = 1
SET(2) = 4
SET(3) = 13
SET(4) = 40
```

To generate a new set involves in most cases just the subtraction of 1 from the last denomination. For example, if ⇒ means 'is followed by':

$$1 \quad 4 \quad 13 \quad 40 \Rightarrow 1 \quad 4 \quad 13 \quad 39$$

When, however, the last value is at its lowest limit (i.e. one more than the second last) then it is the second last which has 1 subtracted from it and the last value is reset to its upper limit (3 * second last + 1).

$$1 \quad 4 \quad 13 \quad 14 \Rightarrow 1 \quad 4 \quad 12 \quad 37$$

Of course, a number of the values may reach their lower limit together:

$$1 \quad 4 \quad 5 \quad 6 \Rightarrow 8 \quad 1 \quad 3 \quad 10 \quad 31$$

and when they have all reached their lower limit there are no further sets.

Having done this, we can combine the procedures into the final program and test it. It will indeed work but it consumes a lot of time. For $N = 4$ there are 331 sets to be tested. It is worth considering whether some of these can be eliminated.

There are two circumstances in which a set can be immediately rejected.

First, if the largest stamp in the set $\leqslant$ best max/3 then clearly it can be rejected. We can at the same time reject others by placing that stamp at its lower limit and selecting the next one.

For example, if best max $= 24$ then:

$$1 \quad 3 \quad 5 \quad 9 \Rightarrow 1 \quad 3 \quad 4 \quad 13$$

We can implement this by adding this test:

```
IF 3 * SET(N) ≤best max THEN
     SET(N) = SET(N−1) + 1
END IF
```

after the generation of the next set.

The second circumstance can best be illustrated by example. Consider:

$$1 \quad 4 \quad 7 \quad 22$$

The max value of this set is 9. Clearly, the largest denomination has no effect, and so we can eliminate 1 4 7 21, 1 4 7 20, etc and consider next 1 4 7 10.

The effect can be promulgated over more than one stamp. For example:

$$1 \quad 4 \quad 13 \quad 40 \Rightarrow 1 \quad 4 \quad 7 \quad 22$$

since the first set has a maximum value of 6.

These two devices reduce the number of sets considered in the $N = 4$ case from 311 to 61, a reduction of approximately five. If we take into account the factor of 4 gained in the procedure for finding the maximum value, we've gained a factor of 20. This clearly illustrates the point that the greatest gains come from strategy rather than tactics.

The Fortran Character Set

Letters: the English alphabet (upper and lower case)
Digits: the Arabic numerals 0 to 9
The underscore: _

The Special Characters:

Blank		:	colon
=	"equals" but really "becomes"	'	apostrophe
+	plus	"	quote
—	minus	;	semi-colon
*	asterisk	!	exclamation mark
/	slash	%	percent
(	open parentheses	&	ampersand
)	close parentheses	<	less than
,	comma	>	greater than
.	decimal point (full stop)	?	question mark
$, £, etc	currency symbol		

Glossary of Terms

Computer programming abounds with jargon. To help the beginner, an explanation is given below of the technical terms used in this book.

Algorithm	A sequence of numerical operations forming the solution to a given mathematical problem.
Alias	In the context of a pointer, a name associated with memory space having an occupant of a different name.
Argument (actual or dummy)	An expression, variable or procedure following in parentheses after a procedure name.
Array	A set of scalars, all of the same type, arranged in orthogonal pattern (rectangle in 2-dimensions, cuboidal in 3-dimensions etc)
Assignment (arithmetic etc)	A statement of the form "variable = expression"
Binary	The number system with base 2 giving the sequence 0, 1, 10, 11, 100 etc
Binding strength	The hierarchy of operators such as $+, -, *$ etc
Bit	Binary digit, namely 1 or 0
Block	A sequence of statements treated as an integral unit which may be embedded in another block or blocks
Branching	The decision process following a choice
Broadcasting	The process by which a scalar object is associated with all the elements of an array
Character	A letter, digit or other symbol
Character set	The particular characters making up a language such as Fortran 90
Character string	A sequence of characters numbered from left to right
Code	Cryptic instructions understood only with the aid of translation (e.g. Machine code, understood

	only by a given machine). Often used loosely to mean "computer program"
Collating sequence	The specific ordering of a given character set
Comment	A "non-executable" part of a program used for descriptive purposes
Compiler	A program, particular to a given computer (processor), which translates a language such as Fortran 90 into machine code
Compound object	An array of data
Concatenation (string)	Chaining together or juxtaposing strings
Conformable (arrays)	Arrays are conformable if they have the same shape
Constant	Data which does not change during execution of a program
Constructor (array)	A means of creating parts of an array by a single statement
Control variable	The variable which determines the number of loop executions
Declare (declaration)	To intimate the names of constants and variables to be used in a program
Default	Assumptions made by the processor in the absence of specific instructions by the programmer
Derived type	A data type made up of a collection of intrinsic or other derived types
Descriptor (format)	Sequence of characters beginning for example with I for INTEGERs which defines the FORMAT of a data item
Dimension (array)	Subscript or number of subscripts, for example A_{ijk} is "three-dimensional"
Dynamic (storage)	Allocation of storage not decided until the program is run
Echo	Printout of what is read in for checking purposes
Element (array)	Single scalar of the set defining an array
Executable statement	An instruction compelling the processor to perform a computation or transfer control
Expression (arithmetic, logical)	A sequence of operands and operators possibly containing parentheses
External (procedure)	A procedure defined outside the compiled Fortran 90 program, possibly in a language other than Fortran 90
Field (width)	Number of characters in a FORMAT
File	A collection of data, characters etc usually external to a program
Fixed point number	A decimal quantity with no exponent part

Floating point number	A decimal quantity with exponent part
Format specification	Sequence of FORMATs
FORTRAN	"FORmula TRANslation" programming languages pre- 1990
Fortran 90	"FORmula TRANSlation" programming language from 1990
Free format	Not under the control of the programmer
Generic name	Name possessed by all intrinsic procedures. Usually does not imply type of argument or result
Identifier	Used synonymously with "name"
Implied (loop)	Loop specified without use of "DO"
Indentation	Program layout involving right-shifted blocks
Initialise	Give starting values to
Input	Information external to a program and requested by the program at run time
Intrinsic	Types, operations, assignments and procedures which are defined in the standard and need no prior specification
Keyword	Reserved word in the language
Label	Numeric pointer to a statement
Language	Formal structure in which programs have to be presented to be understood by the compiler
Library	Collection of (usually) external procedures (usually) in compiled form available to a program
List-directed	Input or output specified only as a list and directed from or to a default device in a default format
Loop (endless loop)	Sequence of instructions obeyed repetitively
Mixed mode	Arithmetic expression involving a mixture of variable TYPEs, for example REAL and INTEGER
Name	A sequence consisting of a letter followed by 0 to 30 alphanumeric characters which stands for a constant or variable
Nesting	The process by which loops are contained within other loop(s)
Operand	An expression that precedes or succeeds an operator
Operator (arithmetic, logical, unary, binary, relational)	A "token" which specifies an operation
Optimise (compiler)	Create machine code which takes best advantage of the facilities of a given processor
Output	Information external to a program created at run time

Parallel computer | A computer capable of performing many tasks simultaneously on different processors

Parameter | Used synonymously with "argument"

PC | "Personal Computer". Currently (1994) of modest capacity (320 Megabytes?) and speed (10 Megaflops?)

Pointer | A single data object which stands for another (a "target"), which may be a compound object such as an array

Portability | The ability for code to be processed by a variety of different processors

Precision | The accuracy with which numbers are stored and manipulated

Precision (double) | The use of twice the default "word length" to increase precision

Procedure | A piece of computation "invoked" or "called" by a program. It may be a FUNCTION or SUBROUTINE and may be "internal" or "external" to the calling program

Processor | A computing system and the means by which programs are made usable on that system

Program (subprogram) | A "main" program is a single program unit or executable program. It may call various procedures as subprograms

Rank (array) | Number of subscripted variables the array has. A scalar has rank zero

Record | A sequence of quantities treated in its entirety as part of a "file"

Section (array) (sub-array) | Part of an array

Shape (array) | The rank of an array and the extent of each subscript

Size (array) | The total number of elements an array has

Source form (free and fixed) | The appearance of a program on screen or page. The Fortran 90 form is "free", although not totally so; the FORTRAN 77 form is "fixed"

Specific name | Name possessed by some intrinsic procedures implying the type of arguments or result

Standard | An agreed international definition for example of a language

Statement | A sequence of "tokens" usually consisting of a single line. It may be continued to new lines using the ampersand or separated from other statements on that line by a semi-colon

Stride | The increment specified for array subscripted variables

String (substring)	A contiguous sequence of characters. (A contiguous part of such a sequence)
Structure chart	A diagram describing pictorially the sequence of operations in a program
Structured programming	An orderly strategy for programming avoiding obscure branching for example
Subscript (subscripted variable)	A scalar INTEGER or INTEGER expression denoting an array dimension. (Element of an array)
Supercomputer	An expensive computer, currently (1994) of capacity several gigabytes and potential speed of teraflops
Target	The data object pointed to by a pointer
Text	Sequences of characters not manipulated numerically
Token (lexical)	A sequence of one or more characters to be interpreted as one, for example **
Truncation	Loss of precision usually in REAL to INTEGER conversions
Truth table	A table describing the results of binary logical operations
Upward-compatibility	The process by which older FORTRANs are recognisable by a Fortran 90 compiler
User-defined	Not intrinsic
Utility routines	In the context of a SUBROUTINE library, those routines which check the characteristics of a given system or processor
Variable	A data object whose value can change during execution of a program
Vector computer	A computer containing specialised components capable of processing a complete vector in the same time as taken for a scalar
Word length	The number of binary bits used by default to represent a (floating point) number

Intrinsic Procedures

An explanation of the arguments A, I, X, STRING etc is given at the end of the list.

Name	Page	Description
ABS (A)		Absolute value.
ACHAR (I)		Character in position I of ASCII collating sequence.
ACOS (X)		Arc cosine (inverse cosine) function.
ADJUSTL (STRING)		Adjust left, removing leading blanks and inserting trailing blanks.
ADJUSTR (STRING)		Adjust right, removing trailing blanks and inserting leading blanks.
AIMAG (Z)		Imaginary part of complex number.
AINT (A [,KIND])		Truncate to a whole number.
ALL (MASK [,DIM])		True if all array elements are true.
ALLOCATED (ARRAY)		True if the array is allocated.
ANINT (A [,KIND])		Nearest whole number.
ANY (MASK [,DIM])		True if any array element is true.
ASIN (X)		Arcsine (inverse sine) function.
ASSOCIATED (POINTER [,TARGET])		True if pointer is associated with target.
ATAN (X)		Arctangent (inverse tangent) function.
ATAN2 (Y,X)		Arctangent for argument of complex number (X, Y).
BIT_SIZE (I)		Maximum number of bits that may be held in an integer.
BTEST (I,POS)		True if bit I of integer POS has value 1.
CEILING (A)		Least integer greater than or equal to its argument.
CHAR (I [,KIND])		Character in position I of the processor collating sequence.
CMPLX (X [,Y] [,KIND])		Convert to COMPLEX type
CONJG (Z)		Conjugate of a complex number.
COS (X)		Cosine function.

COSH (X)	Hyperbolic cosine function.
COUNT (MASK [,DIM])	Number of true array elements.
CSHIFT (ARRAY, SHIFT, [,DIM])	Perform circular shift.
CALL DATE_AND_TIME ([DATE][,TIME] [,ZONE][,VALUES])	Real-time clock reading date and time.
DBLE (A)	Convert to double precision real.
DIGITS (X)	Number of significant digits in the model for X.
DIM (X,Y)	Max (X − Y,0).
DOT_PRODUCT (VECTOR_A, VECTOR_B)	Dot product.
DPROD (X,Y)	Double precision real product of two default real scalars.
EOSHIFT (ARRAY, SHIFT [,BOUNDARY][,DIM])	Perform end-off shift.
EPSILON (X)	Number that is almost negligible compared with one in the model for numbers like X.
EXP (X)	Exponential function.
EXPONENT (X)	Exponent part of the model for X.
FLOOR (A)	Greatest integer less than or equal to its argument.
FRACTION (X)	Fractional part of the model for X.
HUGE (X)	Largest number in the model for numbers like X.
IACHAR (C)	Position of character C in ASCII collating sequence.
IAND (I,J)	Logical AND on the bits.
IBCLR (I,POS)	Clear bit POS to zero.
IBITS (I,POS,LEN)	Extract a sequence of bits.
IBSET (I,POS)	Set bit POS to one.
ICHAR (C)	Position of character C in the processor collating sequence
IEOR (I,J)	Exclusive OR on the bits.
INDEX (STRING, SUBSTRING [,BACK])	Starting position of SUBSTRING within STRING.
INT (A [,KIND])	Convert to integer type.
IOR (I,J)	Inclusive OR on the bits.
ISHFT (I, ISHFT)	Logical shift on the bits.
ISHFTC (I, ISHFT[,SIZE])	Logical circular shift on set of bits on the right.
KIND (X)	Kind type parameter value.
LBOUND (ARRAY [,DIM])	Array lower bounds.
LEN (STRING)	Character length.
LENTRIM (STRING)	Length of STRING without trailing blanks.
LGE (STRING_A, STRING_B)	True if STRING_A equals or follows STRING_B in ASCII collating sequence.
LGT (STRING_A, STRING_B)	True if STRING_A follows STRING_B in ASCII collating sequence.

LLE (STRING_A, STRING_B)	True if STRING_A equals or precedes STRING_B in ASCII collating sequence.
LLT (STRING_A, STRING_B	True if STRING_A precedes STRING_B in ASCII collating sequence.
LOG (X)	Natural (base e) logarithm function.
LOGICAL (L [,KIND])	Convert between kinds of logicals.
LOG10 (X)	Common (base 10) logarithm function.
MATMUL (MATRIX_A, MATRIX_B)	Matrix multiplication.
MAX (A1, A2 [,A3, ...])	Maximum value.
MAXEXPONENT (X)	Maximum exponent in the model for numbers like X.
MAXLOC (ARRAY [,MASK])	Location of maximum array element.
MAXVAL (ARRAY [,DIM] [,MASK])	Value of maximum array element.
MERGE (TSOURCE, FSOURCE, MASK)	TSOURCE when MASK is true and FSOURCE otherwise.
MIN (A1, A2 [,A3, ...])	Minimum value.
MINEXPONENT (X)	Minimum exponent in the model for numbers like X.
MINLOC (ARRAY [,MASK])	Location of minimum array element.
MINVAL (ARRAY [,DIM] [,MASK])	Value of minimum array element.
MOD (A,P)	Remainder modulo P, that is $A - INT(A/P) * P$.
MODULO (A,P)	A modulo P.
CALL MBVITS (FROM, FROMPOS, LEN, TO, TOPOS)	Copy bits.
NEAREST (X,S)	Nearest different machine number in the direction given by the sign S.
NINT (A [,KIND])	Nearest integer.
NOT (I)	Logical complement of the bits.
PACK (ARRAY, MASK [,VECTOR])	Pack elements corresponding to true elements of MASK into rank-one result.
PRECISION (X)	Decimal precision in the model for X.
PRESENT (A)	True if optional argument is present.
PRODUCT (ARRAY [,DIM] [,MASK])	Product of array elements.
RADIX (X)	Base of the model for numbers like X.
CALL RANDOM_NUMBER (HARVEST)	Random numbers in range $0 \leqslant x < 1$.
CALL RANDOM_SEED ([SIZE] [,PUT] [,GET])	Initialise or restart random number generator.
RANGE (X)	Decimal exponent range in the model for X.

REAL (A [,KIND])	Convert to real type.
REPEAT (STRING, NCOPIES)	Concatenates NCOPIES of STRING.
RESHAPE (SOURCE, SHAPE [,PAD] [,ORDER])	Reshape SOURCE to shape SHAPE.
RRSPACING (X)	Reciprocal of the relative spacing of model numbers near X.
SCALE (X,I)	$X \times b^I$, where $b = $ RADIX(X).
SCAN (STRING, SET [,BACK])	Index of left-hand (right-most if BACK is true) character of STRING that belongs to SET; zero if none belong.
SELECTED_INT_KIND (R)	Kind of type parameter for specified exponent range.
SELECTED_REAL_KIND ([P] [,R])	Kind of type parameter for specified precision and exponent range.
SET_EXPONENT (X,I)	Model number whose sign and fractional part are those of X and whose exponent part is I.
SHAPE (SOURCE)	Array (or scalar) shape.
SIGN (A,B)	Absolute value of A times sign of B.
SIN (X)	Sine function.
SINH (X)	Hyperbolic sine function.
SIZE (ARRAY [,DIM])	Array size.
SPACING (X)	Absolute spacing of model numbers near X.
SPREAD (SOURCE, DIM, NCOPIES)	NCOPIES copies of SOURCE forming an array of rank one greater.
SQRT (X)	Square root function.
SUM (ARRAY [,DIM] [,MASK])	Sum of array elements.
CALL SYSTEM_CLOCK ([COUNT] [,COUNT_RATE] [,COUNT_MAX])	Integer data from real-time clock.
TAN (X)	Tangent function.
TANH (X)	Hyperbolic tangent function.
TINY (X)	Smallest positive number in the model for numbers like X.
TRANSFER (SOURCE, MOLD [,SIZE])	Same physical representation as SOURCE, but type of MOLD.
TRANSPOSE (MATRIX)	Matrix transpose.
TRIM (STRING)	Remove trailing blanks from a single string.
UBOUND (ARRAY [,DIM])	Array upper bounds.
UNPACK (VECTOR, MASK, FIELD)	Unpack elements of VECTOR corresponding to true elements of MASK.
VERIFY (STRING, SET [,BACK])	Zero if all characters of STRING belong to SET or index of left-most (right-most if BACK true) that does not.

In the above, the following are used as arguments:

[ ]	Denotes that the argument is optional
A	Numerical argument whose TYPE varies
A1,A2,A3, . . .	Numerical arguments whose TYPE varies
ARRAY	Array argument
B	Numerical argument
BACK	Reverse scan
BOUNDARY	Infilling data
C	Character argument
COUNT	Clock count result
COUNT_RATE	Counts per second result
COUNT_MAX	Maximum number of counts result
DATE	Century, year, month, day result
DIM	Dimension parameter
FIELD	Data array
FROM	Source INTEGER
FROMPOS	Source bit position
FSOURCE	Array option
GET	Re-initialising INTEGERs
HARVEST	REAL argument
I	INTEGER argument
ISHFT	INTEGER result
J	INTEGER argument
KIND	KIND parameter
L	LOGICAL argument
LEN	INTEGER number of bits moved
MASK	Optional LOGICAL array argument
MATRIX	Two-dimensional array argument
MATRIX_A, MATRIX_B	Array arguments
MOLD	Desired data TYPE
NCOPIES	Number of repeated strings
ORDER	Ordering of elements
P	Division or precision parameter
PAD	Array for padding out
POINTER	Argument for association
POS	Position
PUT	Initialisation INTEGERs
R	Range parameter
S	Number directing by its sign
SET	Character string
SHAPE	Rank one INTEGER array
SHIFT	INTEGER number of places
SIZE	Array size
SOURCE	Data array

STRING	Character string
STRING_A, STRING_B	Character strings
SUBSTRING	Character substring
TARGET	Argument to be associated
TIME	Hours, minutes, seconds result
TO	Target INTEGER
TOPOS	Target bit position
TSOURCE	Array option
VALUES	Rank one INTEGER array
VECTOR	Rank one array
VECTOR_A, VECTOR_B	Rank one arrays
X	Numerical argument, usually REAL but can be COMPLEX
Y	Second numerical argument like X
Z	Complex numerical argument
ZONE	Time zone

Options Lists for OPENing and CLOSEing FILES

OPEN (UNIT = unit number, options list)

The full options list contains the specifiers:

IOSTAT =   **ios**, where **ios** is a default integer variable which is set to zero if the statement is correctly executed, and to a positive value otherwise.

ERR =   **error-label**, where **error-label** is the label of a statement to which control will be transferred in the event of an error occurring during execution of the statement.

FILE =   **fln**, where **fln** is a character expression that provides the name of the file. If this specifer is omitted and the unit is not connected to a file, the STATUS = specifier must be specified with the value SCRATCH and the file connected to the unit will then depend on the computer system.

STATUS =   **st**, where **st** is a character expression that provides the value OLD, NEW, REPLACE, SCRATCH, or UNKNOWN. The FILE = specifier must be present if OLD, NEW, or REPLACE is specified; it must not be present if SCRATCH is specified.

If OLD is specified, the file must already exist; if NEW is specified, the file must not already exist, but will be brought into existence by the action of the OPEN statement. The status of the file then becomes OLD.

If REPLACE is specified and the file does not already exist, the file is created; if the file does exist, the file is deleted, and a new file is created with the same name. In each case the status is changed to OLD.

If the value SCRATCH is specified, the file is created and becomes connected, but it cannot be kept after completion of the program or execution of a CLOSE statement.

If UNKNOWN is specified, the status of the file is system dependent. This is the default value of the specifer, if it is omitted.

ACCESS =   **acc**, where **acc** is a character expression that provides the value SEQUENTIAL or DIRECT. For a file which already exists, this value must be an allowed value. If the file does not already exist, it will be brought into existence with the appropriate access method. If this specifier is omitted, the value SEQUENTIAL will be assumed.

FORM =   **fm**, where **fm** is a character expression that provides the value FORMATTED or UNFORMATTED, and determines whether the file is to be connected for formatted or unformatted I/O. For a file which already exists, the value must be an allowed value. If the file does not already exist, it will be brought into existence with an allowed set of forms that includes the specified form. If this specifier is omitted, the default is FORMATTED for sequential access (the usual case) and UNFORMATTED for "direct-access" connection.

RECL =   **rl**, where **rl** is an integer expression whose value must be positive. For a direct-access file, it specifies the length of the records, and is obligatory. For a sequential file, it specifies the maximum length of a record, and is optional with a default value that is processor dependent. For formatted files, the length is the number of characters for records that contain only default characters; for unformatted files it is system dependent but the INQUIRE statement (See below) may be used to find the length of an I/O list. In either case, for a file which already exists, the value specified must be allowed for that file. If the file does not already exist, the file will be brought into existence with an allowed set of record lengths that includes the specified value.

BLANK =   **bl**, where **bl** is a character expression that provides the value NULL or ZERO. This connection must be for formatted I/O.

POSITION =   **pos**, where **pos** is a character expression that provides the value ASIS, REWIND or APPEND. The access method must be sequential, and if the specifier is omitted the default value ASIS will be assumed. A new file is positioned at its initial point. If ASIS is specified and the file exists and is already connected, the file is opened without changing its position; if REWIND is specified the file is positioned at its initial point; if APPEND is specified and the file exists, it is positioned ahead of the endfile record if it has one (and otherwise at its initial point). For a file which exists but is not connected, the effect of the ASIS specifier on the file's position is unspecified.

ACTION =   **act**, where **act** is a character expression that provides the value READ, WRITE, or READWRITE. If READ is specified, the

DELIM=

WRITE, PRINT, and ENDFILE statements must not be used for this connection; if WRITE is specified, the READ statement must not be used; if READWRITE is specified, there is no restriction. If the specifier is omitted, the default value is processor dependent.

del, where **del** is a character expression that provides the value APOSTROPHE, QUOTE or NONE. If APOSTROPHE or QUOTE is specified, the corresponding character will be used to delimit character constants written with list-directed or NAMELIST formatting, and it will be doubled where it appears within such a character constant; also non-default character values will be preceded by kind values. No delimiting character is used if NONE is specified, nor does any doubling take place. NONE is the default value if the specifier is omitted. This specifier may appear only for formatted files.

PAD=

pad, where **pad** is a character expression that provides the value YES or NO. If YES is specified, a formatted input record will be regarded as padded out with blanks whenever an input list and the associated format specify more data than appear in the record. (If NO is specified, the length of the input record must not be less than that specified by the input list and the associated format, except in the presence of an ADVANCE = 'NO' specifier and either an EOR = or an IOSTAT = specification.) The default value if the specifier is omitted is YES. For non-default characters, the blank padding character is processor dependent.

CLOSE (UNIT = unit number, options list)

See Chapter 9 for the short options list

INQUIRE (UNIT = u, options list)

where as usual "UNIT =" is optional, and

INQUIRE (FILE = filename, options list)

The full options list contains the specifiers:

IOSTAT=

ios and ERR = **error-label,** have the meanings described for them in the OPEN statement above. The IOSTAT = variable is the only one which is defined if an error condition occurs during the execution of the statement.

EXIST=

ex, where **ex** is a logical variable. The value .TRUE. is assigned to **ex** if the file (or unit) exists, and .FALSE. otherwise.

OPENED=

open, where **open** is a logical variable. The value .TRUE. is assigned to open if the file (or unit) is connected to a unit (or file), and .FALSE. otherwise.

NUMBER =   **num**, where **num** is an integer variable that is assigned the value of the unit number connected to the file, or -1 if no unit is connected to the file.

NAMED =   **nmd**, and NAME = **nam**, where **nmd** is a logical variable that is assigned the value .TRUE. if the file has a name, and .FALSE. otherwise. If the file has a name, the character variable **nam** will be assigned the name. This value is not necessarily the same as that given in the FILE specifier, if used, but may be qualified in some way. However, in all cases it is a name which is valid for use in a subsequent OPEN statement, and so the INQUIRE can be used to determine the actual name of a file before connecting it. Whether the file name is case sensitive is system dependent.

ACCESS =   **acc**, where **acc** is a character variable that is assigned the value SEQUENTIAL or DIRECT, depending on the access method for a file that is connected, and UNDEFINED if there is no connection.

SEQUENTIAL =   **seq** and DIRECT = **dir**, where **seq** and **dir** are character variables that are assigned the value YES, NO, or UNKNOWN, depending on whether the file **may** be opened for sequential or direct access respectively, or whether this cannot be determined.

FORM =   **frm**, where **frm** is a character variable that is assigned the value FORMATTED or UNFORMATTED, depending on the form for which the file is actually connected, and UNDEFINED if there is no connection.

FORMATTED =   **fmt** and UNFORMATTED = **unf**, where **fmt** and **unf** are character variables that are assigned the value YES, NO, or UNKNOWN, depending on whether the file **may** be opened for formatted or unformatted access, respectively, or whether this cannot be determined.

RECL =   **rec**, where **rec** is an integer variable that is assigned the value of the maximum record length allowed for the file. The length is the number of characters for formatted records containing only characters of default type, and system dependent otherwise. If there is no connection, **rec** becomes undefined.

NEXTREC =   **nr**, where **nr** is an integer variable that is assigned the value of the number of the last record read or written, plus one. If no record has been yet read or written, it is assigned the value 1. If the file is not connected for direct access or if the position is indeterminate because of a previous error, **nr** becomes undefined.

BLANK =   **bl**, where **bl** is a character variable that is assigned the value NULL or ZERO, depending on whether the blanks in numeric fields are by default to be interpreted as null fields or zeros, respectively, and UNDEFINED if there is either no connection. or if the connection is not for formatted I/O.

POSITION=   **pos**, where **pos** is a character variable that is assigned the value REWIND, APPEND, or ASIS, as specified in the corresponding OPEN statement, if the file has not been repositioned since it was opened. If there is no connection, or if the file is connected for direct access the value is UNDEFINED. If the file has been repositioned since the connection was established, the value is processor dependent (but must not be REWIND or APPEND unless that corresponds to the true position).

ACTION=   **act**, where **act** is a character variable that is assigned the value READ, WRITE, or READWRITE, according to the connection. If there is no connection, the value assigned is UNDEFINED.

READ=   **rd**, where **rd** is a character variable that is assigned the value YES, NO or UNKNOWN according to whether READ is allowed, not allowed, or is undetermined for the file.

WRITE=   **wr**, where **wr** is a character variable that is assigned the value YES, NO or UNKNOWN according to whether WRITE is allowed, not allowed, or is undetermined for the file.

READWRITE=   **rw**, where **rw** is a character variable that is assigned the value YES, NO or UNKNOWN according to whether READWRITE is allowed, not allowed, or is undetermined for the file.

DELIM=   **del**, where **del** is a character variable that is assigned the value APOSTROPHE, QUOTE, or NONE, as specified by the corresponding OPEN statement (or by default). If there is no connection, or if the file is not connected for formatted I/O, the value assigned is UNDEFINED.

PAD=   **pad**, where **pad** is a character variable that is assigned the value YES if so specified by the corresponding OPEN statement (or by default), and otherwise NO.

Answers to Exercises

Chapter 2

1. (a), (c), (f) are valid.
   - (b)   does not begin with a letter
   - (d)   contains the illegal character .
   - (e)   contains the illegal character /
   - (g)   contains the illegal character (blank)
   - (h)   contains the illegal characters ( )

   Note.
   There may be confusion between the variable SQRT and the standard FUNCTION
   SQRT (argument) for taking square roots. Its use as a variable name is therefore
   discouraged. Refer to Appendix 3 for a list of the names of procedures such as SQRT,
   EPSILON and so on.

2. The variable D has never been assigned a value and is therefore "undefined". In some
   implementations the program will fail, in others asterisks meaning "undefined" will be
   output.
       Other systems again will take D to be zero, and others will give D the value of the
   variable which occupied that address when a previous program was run. The moral is,
   always give variables the values you want them to have. Never assume anything.
       As a note on style, the program is not commented to say what it is trying to do. The
   variables A, B etc could be given more meaningful names, and should be declared with
   their TYPEs. The STOP instruction is superfluous and it helps to add "PROGRAM
   TEST" after "END" as a form of commenting to help the user.
       IMPLICIT NONE should always be specified.

Chapter 3

1. (a) All variables are of TYPE REAL and so conversions will take place leading to P,Q and
       R being (very close to) 5.0, −9.0 and 4.0 respectively. Therefore X,Y and Z will all
       have values (very close to) −11.25.

   (b) INTEGER expressions would be truncated before assignment to the REAL
       variables X,Y,Z which would take the values (very close to) −11.0, −10.0 and −9.0
       respectively.

2. In the first version all quantities are REAL and no conflict arises. In the second, although the INTEGERs 9, 5 and 32 appear, the presence of the REAL variable CENTIGRADE means that 9 and 5 will be converted to REAL equivalents before the multiplication and division are done. Similarly 32 will be converted to its REAL equivalent before the addition is done, and FAHRENHEIT will have a value very close to that found in version one. In the third version, however, the result of the INTEGER divide 9/5 is INTEGER and will be rounded to 1 before conversion and multiplication by REAL CENTIGRADE to give a REAL result. On a typical run, the versions yielded FAHRENHEIT equivalents of 16°C of 60.8°F, 60.8°F and 48.0°F. A simple program is given below.

```
PROGRAM TEMPERATURE
! convert centigrade to fahrenheit
IMPLICIT NONE
REAL :: FAHRENHEIT, CENTIGRADE
READ*, CENTRIGRADE
  FAHRENHEIT = 1.8*CENTIGRADE + 32.0; PRINT*, FAHRENHEIT
  FAHRENHEIT = 9*CENTIGRADE/5 + 32; PRINT*, FAHRENHEIT
  FAHRENHEIT = 9/5*CENTIGRADE + 32; PRINT*, FAHRENHEIT
END PROGRAM TEMPERATURE
```

3. The resulting values of X,Y,M,N are 30.0, 17.5, −5 and 16 respectively.

4. (a) $X + Y*Y*Y$ or $X + Y**3$
   (b) $(X+Y)**3$
   (c) $X**4$ or $(X**2)**2$
   (d) $(A+B)/C$
   (e) $(A+B)/(C+D)$
   (f) $(A+B)/(C+D/(E+F))$
   (g) $1. + X + X**2*(1. + X/3.)/2.$

5. (a), (b), (d) are valid.
   (c) contains the illegal character ,
   (e) contains the illegal character *
   (f) contains the illegal character (blank)
   (g) contains the illegal character ±

6. In due course (Chapter 5) programs will become complicated enough for us to like to do a bit of preliminary work on the "structure" of the program before beginning the coding. Even in this simple example, note that the variable S occurs several times, as does the algebra $(S−A)(S−B)(S−C)$. Although S first appears as "2S" note that this cannot be used as a Fortran 90 name. Resist the temptation to use IR as the name of the inscribed radius. It will usually be a REAL quantity.

   A possible program structure might be:

   > Read in A,B and C
   > Echo the input as a check
   > Calculate S, which is $(A+B+C)/2.0$
   > Calculate $(S−A)*(S−B)*(S−C)$ and call it SMINUS
   > Calculate and print the inscribed and
   >   escribed radii

and in Fortran 90:

```
    PROGRAM CIRCLES
! for a given triangle ABC this program reads in the sides A,B and
! C and calculates the radii of the inscribed and escribed circles
    IMPLICIT NONE
    REAL :: A,B,C,S,SMINUS
    PRINT*, 'PLEASE TYPE IN THE SIDES OF THE TRIANGLE &
            IN THE ORDER A,B,C'
    READ*,A,B,C
    PRINT*, 'THE TRIANGLE SPECIFIED HAS SIDES A=',A, 'B=',B, &
            'C=',C
    S=(A+B+C)*0.5; SMINUS=(S-A)*(S-B)*(S-C)
    PRINT*, 'THE RADIUS OF THE INSCRIBED CIRCLE IS', SQRT &
            (SMINUS/S)
    PRINT*, 'THE RADIUS OF THE ESCRIBED CIRCLE IS', A*B*C/ &
            (4.0*SQRT (S*SMINUS))
    END PROGRAM CIRCLES
```

For A, B and C of 1.0 the inscribed radius is 0.2886751 and the escribed radius is 0.5773502. For A,B and C having values of 65.23, 39.77 and 58.05 respectively the inscribed radius is 13.9971247 and the escribed radius is 32.9925270.

You may well also be advised to guard against impossible triangles.

Chapter 4

1.
```
    PROGRAM BEAM_ON_ELASTIC_FOUNDATION
! to calculate the deflection of a loaded beam resting on
! an elastic foundation made of springs
    IMPLICIT NONE
    REAL :: LENGTH, K, EI, X, P, BETA, BETAL, BETAX, BETALX, SINHBL, &
            SINBL, CALC, V
    PRINT*, 'LENGTH OF BEAM IN METRES?'
    READ*, LENGTH
    PRINT*, 'BEAM FLEXURAL RIGIDITY IN KN/METRE SQUARED?'
    READ*, EI
    PRINT*, 'DISTANCE OF DEFLECTION POINT FROM LOAD IN METRES?'
    READ*, X
    PRINT*, 'MAGNITUDE OF FORCE IN KILONEWTONS?'
    READ*, P
    PRINT*, 'FOUNDATION STIFFNESS IN KN/METRE/METRE'
    READ*, K
      BETA=SQRT (SQRT (K/(4.0*EI))); BETAL=BETA*LENGTH
      BETAX=BETA*X; BETALX=BETA* (LENGTH-X)
      SINHBL=SINH (BETAL); SINBL=SIN (BETAL)
      CALC=SINHBL*COS (BETAX)*COSH (BETALX)-SINBL*COSH (BETAX)* &
            COS (BETALX)
      V=2.0*P*BETA*CALC/ (K* (SINHBL**2-SINBL**2))
    PRINT*, 'THE DEFLECTION OF THE BEAM OF LENGTH', LENGTH, 'METRES'
    PRINT*, 'WITH FLEXURAL RIGIDITY', EI, 'KN*M**2, FORCE', P, &
            'KILONEWTONS'
    PRINT*, 'FOUNDATION STIFFNESS', K, 'KN PER METRE PER METRE'
    PRINT*, 'AT', X, 'METRES FROM THE FORCE IS', V, 'METRES'
    END PROGRAM BEAM_ON_ELASTIC_FOUNDATION
```

For a length of beam of 10, EI $= 10^6$, P $= 100$, K $= 4 \times 10^3$ the resulting deflections are

x = 0   v = 0.010892
x = 2   v = 0.0070077
x = 5   v = 0.0020671

2. In this problem, a lot of the scope for error lies in mixing up the units of the various quantities. We therefore take care to annotate the program accordingly.

```
    PROGRAM SATELLITE_ORBITS
! this program calculates the circular velocity, escape velocity and
! time
! for a complete orbit for a satellite flying at a height of h metres
! above the earth
    IMPLICIT NONE
    REAL : : R, H, VC, VE, PI, ORBIT, TIME
    PRINT*, "PLEASE TYPE IN THE SATELLITE'S HEIGHT ABOVE THE EARTH &
            IN METRES"
    READ*, H
    R = 6271200.0
! units of r and h are metres
    VC = 7749.0*SQRT (R)/SQRT (R + H)*60.0
! units of vc are metres per minute
    VE = VC*SQRT(2.0)/1000.0*60.0
! units of ve are kilometres per hour
    PI = ACOS(-1.0); ORBIT = 2.0*PI*(R + H)
  ! units of orbit are metres
    TIME = ORBIT/VC
! units of time are minutes
    PRINT*, 'THE ESCAPE VELOCITY IS', VE, 'KILOMETRES PER HOUR'
    PRINT*, 'TIME TAKEN FOR ONE ORBIT IS', TIME, 'MINUTES'
END PROGRAM SATELLITE_ORBITS
```

For H of 15 000 m the escape velocity is 39 400 km/hour and the time taken for one orbit is 85 minutes.

3.
```
PROGRAM IMAGINARY_COSINE
! to use intrinsic functions to get cos(x)
IMPLICIT NONE
COMPLEX : : Z
REAL : : PI
PI = 4.*ATAN (1.)
Z = CMPLX (0.0, 0.25*PI)
PRINT*, 'The Cosine of 45 degrees is', (EXP(Z) + 1./EXP(Z))*0.5
END PROGRAM IMAGINARY_COSINE
```

The cosine of 45 degrees is (0.7071,0.)

Chapter 5

1.

```
            read n
            sum = 0.
            sum_of_squares = 0.
  large_store = 0.;small_store = 1000000.0
        ┌─────────────────────────────────┐
        │  n times                        │
        ├─────────────────────────────────┤
        │   read x                        │
        │   largest = max (x, large_store)│
        │   smallest = min (x, small_store)│
        │                                 │
        │   large_store = largest         │
        │   small_store = smallest        │
        │                                 │
        │     add x to sum                │
        │     add x² to sum_of_squares    │
        └─────────────────────────────────┘
      print sum/n
```

$$\text{print sqrt}\left(\frac{\text{sum\_of\_squares} - \text{sum}^2/n}{n - 1}\right)$$

print largest, smallest, largest-smallest

```
PROGRAM STATISTICS
! a program to calculate simple statistics for an input list of
! numbers
  IMPLICIT NONE
  INTEGER :: N, I; REAL :: X, LARGEST, SMALLEST
  REAL :: SUM=0.0, SUM_OF_SQUARES=0.0, LARGE_STORE=0.0, SMALL_ &
    & STORE = 1000000.0
  READ*, N
  DO I = 1, N
    READ*,X
    LARGEST=MAX(X, LARGE_STORE); SMALLEST=MIN(X, SMALL_STORE)
    LARGE_STORE = LARGEST; SMALL_STORE = SMALLEST
    SUM = SUM + X
    SUM_OF_SQUARES = SUM_OF_SQUARES + X*X
  END DO
  PRINT*, 'THE MEAN IS', SUM/N
  PRINT*, 'THE STANDARD DEVIATION IS', SQRT((SUM_OF_SQUARES- &
      SUM**2/N)/(N-1))
  PRINT*, 'THE LARGEST AND SMALLEST NUMBERS ARE', LARGE_STORE, &
      'AND', SMALL_STORE, 'RESPECTIVELY'
  PRINT*, 'THE RANGE IS', LARGE_STORE-SMALL_STORE
END PROGRAM STATISTICS
```

(a) 10.42 (b) 3.4627 (c) 15.23 (d) 5.67 (e) 9.56

2.

$$\text{read a,b,n}$$

$$\text{calculate f1} = \frac{\sqrt{a}\sin a}{a + e^a}$$

$$\text{interval} = \frac{b - a}{n}$$

$$\text{integral} = 0.$$

n times
$y = a + \text{interval}$
calculate $f2 = \dfrac{\sqrt{y}\sin y}{y + e^y}$
add $\dfrac{f1 + f2}{2}$ (interval) to integral
set f1 to f2 repeat

print out integral

And in Fortran 90:

```
PROGRAM TRAPEZIUM_RULE
! a program to evaluate integrals using the trapezium rule
  IMPLICIT NONE
    INTEGER:: N, I
    REAL :: A, B, F1, F2, INTERVAL, INTEGRAL = 0.0, PI, Y
! the limits of integration a and b are given in degrees
    READ*, A, B, N
! convert to radian measure
    PI = 4.0*ATAN(1.0); A=A*PI/180.0; B=B*PI/180.0
    F1 = SQRT(A)*SIN(A)/(A+EXP(A))
    INTERVAL = (B-A)/N
    Y = A
      DO I = 1, N
        Y = Y + INTERVAL
        F2 = SQRT(Y)*SIN(Y)/(Y+EXP(Y))
        INTEGRAL = INTEGRAL + (F1+F2)/2.0*INTERVAL
        F1 = F2
      END DO
    PRINT*, 'THE INTEGRAL IS', INTEGRAL
END PROGRAM TRAPEZIUM_RULE
```

For 10 steps the integral is 0.2642579. For 100 steps it is 0.2647655.

3. First note that, with the exception of the leading term 1, each term of the cosine series is found by multiplying the preceding term by $-x^2$ and dividing by $2j(2j-1)$ where

$j = 1,2,3,\ldots$ number of terms wanted. Thus a viable structure chart might be

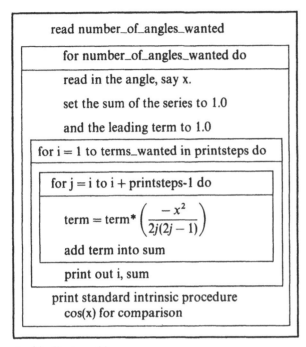

And in Fortran 90:

```
  PROGRAM COSINE_SERIES
! a program to sum the series for cos(x) to n terms
  IMPLICIT NONE
  INTEGER :: NUMBER_OF_ANGLES_WANTED, I, J, K, TERMS_WANTED, &
    PRINTSTEPS
    REAL :: ANGLE, SUM, TERM, PI
      PI = ACOS(-1.0)
      READ*, NUMBER_OF_ANGLES_WANTED
        DO K = 1, NUMBER_OF_ANGLES_WANTED
          READ*, ANGLE, TERMS_WANTED, PRINTSTEPS
          ANGLE = ANGLE*PI/180.0
          SUM = 1.0;TERM = 1.0
          DO I = 1, TERMS_WANTED, PRINTSTEPS
            DO J = I, I + PRINTSTEPS - 1
              TERM = TERM*(-ANGLE**2/(2*J*(2*J-1)))
              SUM = SUM + TERM
            END DO
            PRINT*, I, SUM
          END DO
          PRINT*, 'THE INTRINSIC VALUE IS', COS(ANGLE)
        END DO
END PROGRAM COSINE_SERIES
```

For large values of x, the series solution implies differences between very large quantities. For example a typical run of COSINE_SERIES produced the results:

x°	Solution after 50 terms	Intrinsic solution
30	0.8660254	0.8660254
570	-0.8661471	-0.8660254
1470	$-4.2664737E+02$	0.8660246

The intrinsic function routine probably converts large x to an equivalent smaller x before summing a series, not necessarily this one.

Chapter 6

1.

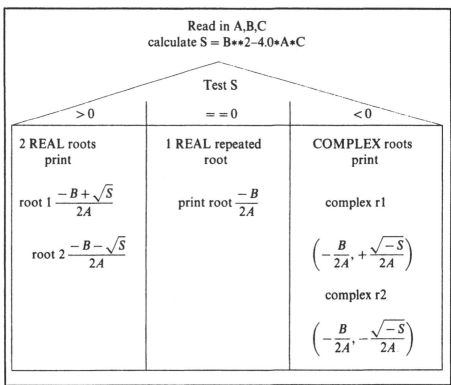

In Fortran 90:

```
PROGRAM CALCULATE_ROOTS
! to calculate the roots of a quadratic
IMPLICIT NONE
REAL :: A, B, C, S, REAL_ROOT1, REAL_ROOT2, ROOTS
COMPLEX :: COMPLEX_ROOT1, COMPLEX_ROOT2
INTEGER :: I, N
READ*, N
  DO I = 1, N
    READ*, A, B, C
    S = B**2-4.0*A*C
```

```
IF (S>0.0) THEN
   PRINT*, 'THE ROOTS ARE REAL AND UNEQUAL'
   ROOTS=SQRT (S)
   REAL_ROOT1 = (-B+ROOTS)/A*0.5; REAL_ROOT2 = (-B-ROOTS)/ &
             A*0.5
   PRINT*, 'THEIR VALUES ARE', REAL_ROOT1, 'AND', REAL_ROOT2
ELSE IF (S<0.0) THEN
   PRINT*, 'THE ROOTS ARE COMPLEX'
   ROOTS = SQRT(-S)
   COMPLEX_ROOT1 = CMPLX (-B/A*0.5, ROOTS/A*0.5)
   COMPLEX_ROOT2 = CMPLX(-B/A*0.5, -ROOTS/A*0.5)
   PRINT*, 'THEIR VALUES ARE', COMPLEX_ROOT1, 'AND', &
             COMPLEX_ROOT2
ELSE
   PRINT*, 'THE ROOTS ARE REAL AND EQUAL'
   PRINT*, 'THEIR VALUE IS', -B/A*0.5
ENDIF
END DO
END PROGRAM CALCULATE_ROOTS
```

Alternatively, one could write a program which assumes a priori that the roots are COMPLEX, for example:

```
PROGRAM COMPLEX_ROOTS
! a version assuming complex results
IMPLICIT NONE
REAL :: A, B, C; INTEGER :: I, N; COMPLEX :: ROOTS, CA, CB, CC
READ*, N
   DO I = 1, N
     READ*, A, B, C
! make sure a, b, c are converted to TYPE COMPLEX
     CA = COMPLEX(A); CB=CMPLX(B); CC=CMPLX(C)
     ROOTS=SQRT (CB*CB-4.*CA*CC)
     PRINT*, 'THE FIRST ROOT IS', (-CB+ROOTS)/2.0/CA
     PRINT*, 'THE SECOND ROOT IS', (-CB-ROOTS)/2.0/CA
   END DO
END PROGRAM COMPLEX_ROOTS
```

For $A = 1.0$, $B = 2.0$, $C = 3.0$, the roots are $(-1.0, 1.4142)$, $(-1.0, -1.4142)$.

2.

Read in N Calculate R, the INTEGER part of $\left(\dfrac{\sqrt{N}+1}{2}\right)$			
Test N			
$2R(2R-1) \geqslant N \geqslant (2R-1)^2$?	$< 2R^2$?	$< 2R(2R+2)$?	$> 2R(2R+2)$
X is R Y is $N-2R(2R-1)+R$	X is $N-2R^2+R$ Y is R	X is $-R$ Y is $2R(2R+1)-R-N$	X is $N-2R+R$ Y is $-R$
Print out found X and Y			

In Fortran 90:

```
PROGRAM NUMBER_SPIRAL
! a program to locate a number in a spiral
IMPLICIT NONE
INTEGER :: NUMBER, R, X, Y, S, SP1, SM1, S2
REAL :: REAL_NUMBER
READ*, NUMBER
REAL_NUMBER = REAL (NUMBER)
R = INT ( (SQRT (REAL_NUMBER) + 1) /2)
PRINT*, R
S = 2*R; SP1 = S+1; SM1 = S-1; S2 = S*(S+2)
IF (NUMBER > = SM1**2.AND.NUMBER < = S*SM1) THEN
  X = R
  Y = NUMBER - S*SM1 + R
ELSE IF (NUMBER > S*SM1.AND.NUMBER < = S**2) THEN
  Y = R
  X = NUMBER - S**2 + R
ELSE IF NUMBER > S**2.AND.NUMBER < = S*SP1) THEN
  X = -R
  Y = S*SP1 - R - NUMBER
ELSE
  X = NUMBER - S2 + R
  Y = -R
ENDIF
PRINT*, 'THE NUMBER IS LOCATED AT X = ', X, 'Y = ', Y
END PROGRAM NUMBER_SPIRAL
```

625 is found at X = 13, Y = −12.

3.

In Fortran 90:

```
PROGRAM GALLUP_POLL
! to conduct a Gallup poll survey
IMPLICIT NONE
INTEGER :: SAMPLE, I, COUNT, THIS_TIME, LAST_TIME, TOT_REP, TOT_ &
           MAV, TOT_DEM, TOT_OTHER, REP_TO_MAV, DEM_TO_MAV, &
           CHANGED_MIND
READ*, SAMPLE
COUNT = 0; TOT_REP = 0; TOT_MAV = 0; TOT_DEM = 0; TOT_OTHER = 0; REP_ &
      TO_MAV = 0
DEM_TO_MAV = 0; CHANGED_MIND = 0
OPEN (10, FILE = 'GALLUP.DAT')
  DO I = 1, SAMPLE
    COUNT = COUNT + 1
    READ (10, '(I3,I2)', ADVANCE = 'NO') THIS_TIME, LAST_TIME
    VOTES : SELECT CASE (THIS_TIME)
      CASE (1); TOT_REP = TOT_REP + 1
      CASE (3); TOT_MAV = TOT_MAV + 1
        IF (LAST_TIME/ = 3) THEN
          CHANGED_MIND = CHANGED_MIND + 1
          IF (LAST_TIME = = 1) REP_TO_MAV = REP_TO_MAV + 1
          IF (LAST_TIME = = 2) DEM_TO_MAV = DEM_TO_MAV + 1
        END IF
      CASE (2); TOT_DEM = TOT_DEM + 1
      CASE DEFAULT; TOT_OTHER = TOT_OTHER + 1
    END SELECT VOTES
  END DO
    PRINT*, 'Percent  Republican  is',  REAL  (TOT_REP)/REAL  &
          (COUNT)*100.0
    PRINT*, 'Percent Maverick is', REAL (TOT_MAV)/REAL(COUNT)*100.0
    PRINT*, 'Percent Democrat is', REAL (TOT_DEM)/REAL (COUNT)*100.0
    PRINT*, 'Percent Others is', REAL (TOT_OTHER)/REAL (COUNT)*100.0
    PRINT, 'Percent changing Rep to Mav is', REAL (REP_TO_MAV)/&
          REAL (CHANGED_MIND)*100.0
    PRINT, 'Percent changing Dem to Mav is', REAL (DEM_TO_MAV)/&
          REAL (CHANGED_MIND)*100.0
END PROGRAM GALLUP_POLL
```

   Percent Republican 41.67
   Percent Maverick 18.06
   Percent Democrat 29.17
   Percent Others 11.11
   Republican to Maverick 52.94
   Democrat to Maverick 41.18

4. Data for the problem consists of the dam geometry, given by l_1, l_2, h_1 and h_2 together with the height of water y_1.

   Physical parameters are the weight of concrete ($24 \, kN/m^3$), the weight of water (typically $10 \, kN/m^3$) and the coefficient of sliding friction, typically 0.4

> read in geometry and physical data.
> Calculate the weights of dam regions 1,2,3,
> say W_1, W_2, W_3. Calculate external water
> forces F and uplift force U.
> Calculate distances from 0 of points of
> application of these 5 forces.

```
PROGRAM DAM
! check the dam for sliding or toppling
IMPLICIT NONE
REAL :: L1, L2, LEV1, LEV2, LEV3, N, MU, H1, H2, Y1, GAMAC, GAMAW, &
        F, U, W1, W2, W3, FLEV, ULEV, X, FACTOR
LOGICAL :: SLIDE, TOPPLE
PRINT*, 'READ IN GEOMETRIC DATA:L1, L2, H1, H2, Y1'
READ*, L1, L2, H1, H2, Y1
PRINT*, 'READ IN THE PHYSICAL DATA:GAMAC, GAMAW, MU'
READ*, GAMAC, GAMAW, MU
! calculate the forces and ''lever arms'' (points of application
! relative to 0)
W1 = (L1*(H1 + H2))*GAMAC
W2 = (0.5*H2*L2)*GAMAC
W3 = (H1*L2)*GAMAC
F = 0.5*GAMAW*Y1*Y1
U = 0.5*(L1 + L2)*Y1*GAMAW
N = W1 + W2 + W3 - U
LEV1 = L1/2.0
LEV2 = L1 + L2/3.0
LEV3 = L1 + L2/2.0
FLEV = Y1/3.0
ULEV = (L1 + L2)/3.0
! now check stability
SLIDE = .TRUE.
TOPPLE = .TRUE.
IF (F < (N*MU)) SLIDE = .FALSE.
X = (W1*LEV1 + W2*LEV2 + W3*LEV3 + F*FLEV - U*ULEV)/N
IF (X >= (L1 + L2)/3.0.AND.X <= 2.0*(L1 + L2)/3.0) TOPPLE = .FALSE.
IF (.NOT.SLIDE.AND.NOT.TOPPLE) PRINT*, 'THE DAM IS SAFE'
IF (SLIDE.OR.TOPPLE) THEN
  IF (SLIDE) THEN
    PRINT*, 'THE DAM WILL SLIDE'
    ELSE
    PRINT*, 'THE DAM WILL TOPPLE'
  ENDIF
ENDIF
IF (SLIDE.AND.TOPPLE) PRINT*, 'THE DAM WILL SLIDE AND TOPPLE'
FACTOR = N*MU/F
PRINT*, 'THE FACTOR OF SAFETY AGAINST SLIDING IS', FACTOR
PRINT*, 'AND THE POSITION OF THE NORMAL FORCE IS', X
PRINT*, 'METRES FROM THE UPSTREAM FACE'
END PROGRAM DAM
```

The dam is safe with a factor of safety against sliding of 1.7.

5.

```
read in number_of_angles_wanted

for number_of_angles_wanted do

read in the angle, say x
print out x as a check
Set the sum of the series to 0.0
and the leading term to 1.0
Set loop-counter to 0

do

loop-counter = loop-counter + 1

term = term* ( −x² / 2i(2i − 1) )

add term into sum
if absolute value of term is small enough
or if loop-counter reaches limit
then exit

print out loop-counter, sum

print standard intrinsic procedure cos(x)
for comparison
```

$$\text{term} = \text{term} * \left(\frac{-x^2}{2i(2i-1)} \right)$$

And in Fortran 90:

```
PROGRAM COSINE_SERIES_2

! a program to sum the series for cos(x) to a required accuracy
  IMPLICIT NONE
  INTEGER :: NUMBER_OF_ANGLES_WANTED, J, K, PRINTSTEPS, LOOP_LIMIT
  REAL :: ANGLE, SUM, TERM, PI
    PI = ACOS(−1.0)
    READ*, NUMBER_OF_ANGLES WANTED, LOOP_LIMIT
      DO K=1, NUMBER_OF_ANGLES_WANTED
        READ*, ANGLE, PRINTSTEPS
        ANGLE = ANGLE*PI/180.0
        SUM = 1.0; TERM = 1.0; J = 0
        DO
          J = J+1
          TERM = TERM*(−ANGLE**2/(2*J*(2*J−1)))
          SUM = SUM+TERM
          IF (ABS(TERM) < 1.E−6.OR.J = =LOOP_LIMIT) EXIT
          IF (J/PRINTSTEPS*PRINTSTEPS = =J) PRINT*, J, SUM
        END DO
        IF (J = =LOOP_LIMIT) PRINT*, 'WARNING − LOOP LIMIT REACHED'
        PRINT*, 'THE CONVERGED VALUE OF COS (',ANGLE,') RADIANS IS', &
                SUM
        PRINT*, 'IT TOOK', J, 'ITERATIONS TO CONVERGE'
        PRINT*, 'THE INTRINSIC VALUE IS', COS (ANGLE)
      END DO
  END PROGRAM COSINE_SERIES_2
```

cos (570°) converged to −0.8661470 in 18 iterations.

Chapter 7

1.

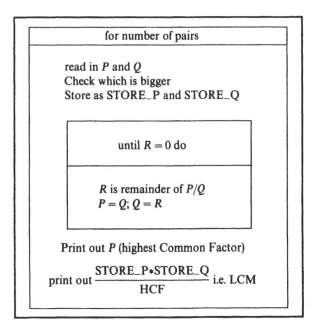

```
PROGRAM HCF_AND_LCM
! a program to calculate the highest common factor
! and lowest common multiple of two integers
IMPLICIT NONE
INTEGER :: P,Q,BIGGER,SMALLER,BIG_STORE,SMALL_STORE,WHOLE, &
         REMAINDER
READ*,P,Q
IF (P>Q) THEN
  BIGGER=P;SMALLER=Q
ELSE
  BIGGER=Q;SMALLER=P
ENDIF
BIG_STORE=BIGGER;SMALL_STORE=SMALLER
    DO
      WHOLE=BIGGER/SMALLER
      REMAINDER=BIGGER-SMALLER*WHOLE
      IF (REMAINDER==0) EXIT
      BIGGER=SMALLER;SMALLER=REMAINDER
    END DO
  PRINT*, 'THE HIGHEST COMMON FACTOR IS', SMALLER
  PRINT*, 'THE LOWEST COMMON MULTIPLE IS', BIG_STORE* &
         SMALL_STORE/SMALLER
END PROGRAM HCF_AND_LCM
```

38, 2660 and 1, 24 respectively.

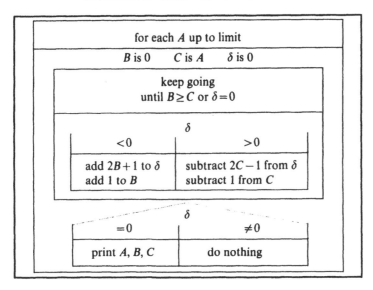

2. PROGRAM PYTHAGOREAN_TRIANGLE

```
PROGRAM PYTHAGOREAN_TRIANGLE
! a program to find all the Pythagorean triangles up to a given limit
IMPLICIT NONE
INTEGER::A,B,C,DELTA,LIMIT,COUNT,TRIANGLES
READ*,LIMIT
TRIANGLES=0
DO A=5,LIMIT
! although delta is really 0 for the trivial case set it to 1
  B=0; C=A; DELTA=1; COUNT=0
  DO
    COUNT=COUNT+1
    IF (DELTA<0) THEN
      DELTA=DELTA+2*B+1
      B=B+1
    ELSE IF (DELTA>0) THEN
! remember to return delta to 0 first time round
      IF (COUNT==1) DELTA=DELTA-1
      DELTA=DELTA-(2*C-1)
      C=C-1
    ENDIF
    IF (B>=C.OR.DELTA==0) EXIT
  END DO
  IF (DELTA==0) THEN
    TRIANGLES=TRANGLES+1
    PRINT*,A,B,C
  ENDIF
END DO
PRINT*,'THERE ARE',TRIANGLES,'PYTHAGOREAN TRIANGLES UP TO SIDE',&
  LIMIT
END PROGRAM PYTHAGOREAN_TRIANGLE
```

There are 18 Pythagorean triangles up to side 50 and 42 up to side 100.

```
3. PROGRAM MORTGAGE_REPAYMENT
   ! a program to calculate monthly repayments
   IMPLICIT NONE
   INTEGER :: PERIOD, PAGE = 0
   REAL :: LOW_RATE, HIGH_RATE, CURRENT_RATE, RATE_INCREMENT, &
           LOW_AMOUNT, HIGH_AMOUNT, CURRENT_AMOUNT, &
           AMOUNT_INCREMENT, ANNUAL_PAYMENT, &
           MONTHLY_PAYMENT, FACTOR
   READ*, PERIOD ! whole number of years
   READ*, LOW_RATE, HIGH_RATE, RATE_INCREMENT ! To nearest .1%
   READ*, LOW_AMOUNT, HIGH_AMOUNT, AMOUNT_INCREMENT ! To nearest 100
        ! pounds
     CURRENT_RATE = LOW_RATE - RATE_INCREMENT
       DO
         PAGE = PAGE + 1; CURRENT_RATE = CURRENT_RATE + RATE_INCREMENT
         PRINT'(/,A,/,A,I3,/,A,I3,A,/,A,/,A,F5.2,A,/,A,/,A,/,A)', &
              '*************************************************', &
              '    MORTGAGE REPAYMENT TABLE PAGE', PAGE, &
              '    (REPAYMENT PERIOD = ', PERIOD, 'YEARS)', &
              '*************************************************', &
              '    ANNUAL INTEREST RATE = ', CURRENT_RATE, '%', &
              '*************************************************', &
              'AMOUNT BORROWED    MONTHLY REPAYMENT', &
              '------ --------    ------- ---------'
         CURRENT_AMOUNT = LOW_AMOUNT - AMOUNT_INCREMENT
         DO
           CURRENT_AMOUNT = CURRENT_AMOUNT + AMOUNT_INCREMENT
           FACTOR = (1. + CURRENT_RATE/100.)**PERIOD
           ANNUAL_PAYMENT = CURRENT_AMOUNT* &
                        CURRENT_RATE*FACTOR/(100.*(FACTOR - 1.))
           MONTHLY PAYMENT = ANNUAL_PAYMENT/12.
           PRINT'(A,F9.2,A,F9.2)', &
              '     ', CURRENT_AMOUNT, '     ', MONTHLY_PAYMENT
           IF (ABS(CURRENT_AMOUNT_HIGH_AMOUNT) < 99.) EXIT
         END DO
         IF ABS(CURRENT_RATE_HIGH_RATE) < .1) EXIT
       END DO
   END PROGRAM MORTGAGE_REPAYMENT
```

For example, $30000 borrowed over 25 years at 5% per annum costs $177.38 per month.

Since "Structure Charts", "Flow Diagrams" and other ways of illustrating essential program structure are so much a matter of personal taste, only a selection of structure charts is given for exercises from now on.

```
4. PROGRAM CUBES_OF_DIGITS
   !
   ! A program to find all 3-digit numbers equal to the sum of their digits
   ! cubed
   IMPLICIT NONE
     INTEGER :: NUMBER, SUM, WORK
```

```
    DO NUMBER=1, 999
      SUM=0
      WORK=NUMBER
      DO
        IF (WORK==0) EXIT
        SUM=SUM+MOD (WORK,10)**3
        WORK=WORK/10
      END DO
      IF (SUM==NUMBER) PRINT*, NUMBER
    END DO
  END PROGRAM CUBES_OF_DIGITS
```

The numbers are:

  1
  153
  370
  371
  407

5.
```
PROGRAM MANDEL
  ! Prints a typical 'Mandelbrot Set'
  IMPLICIT NONE
  CHARACTER :: LINE*80, TEXT*18= ' -.+xXo00aaaa'
  COMPLEX :: Z, C; INTEGER :: I, J, K, KK
  DO J=1, 24; ILOOP: DO I=1, 80
    C=CMPLX(I/25. -2.0, J/12.-1.); Z=0.
    DO K=1, 63
      Z=Z*Z+C; IF (ABS(Z)>2.) EXIT
    END DO
    KK=1+LOG(1.*K)/LOG(2.)
  ! KK=1+K/8
    LINE (I:I) =TEXT(KK:KK); END DO ILOOP
    WRITE (*,'(A)') LINE (1:79)
  END DO
END PROGRAM MANDEL
```

Typical fractal paterns are shown in the following:

```
----------------------------------------------..+..-----------------------------
---------------------------------------------+o+aao+X.---------------------------
---------------------------------------------..aaaaaX.---------------------------
-------------------------------------.+++...+..++aaa+...++.--.-------------------
---------------------------------..+aa+aaaaaaaaaaaaaaax.aax+.--------------------
----------------------------------.+.+aaaaaaaaaaaaaaaaaaax.----------------------
---------------++-----------...aaaaaaaaaaaaaaaaaaaaaaaaaox..---------------------
--------------.......X........aaaaaaaaaaaaaaaaaaaaaaaaaaaa+-----------------------
------------...aaaaaaa+...Xaaaaaaaaaaaaaaaaaaaaaaaaaaa.+--------------------------
-------------....XXaaaaaaaaaa+aaaaaaaaaaaaaaaaaaaaaaaaaaa.------------------------
------------..+X+aaaaaaaaaaaaoaaaaaaaaaaaaaaaaaaaaaaaaaaa.-----------------------
aaaaaaaaaaaaaaaaaaaaaaaaaaaaaaaaaaaaaaaaaaaaaaaaaaaaaaaaaaa+..-------------------
------------..+X+aaaaaaaaaaaaoaaaaaaaaaaaaaaaaaaaaaaaaaaa.-----------------------
-------------....XXaaaaaaaaaa+aaaaaaaaaaaaaaaaaaaaaaaaaaa.------------------------
------------...aaaaaaa+...Xaaaaaaaaaaaaaaaaaaaaaaaaaaa.+--------------------------
--------------.......X........aaaaaaaaaaaaaaaaaaaaaaaaaaaa+-----------------------
---------------++-----------...aaaaaaaaaaaaaaaaaaaaaaaaaox..---------------------
----------------------------------.+.+aaaaaaaaaaaaaaaaaaax.----------------------
---------------------------------..+aa+aaaaaaaaaaaaaaax.aax+.--------------------
-------------------------------------.+++...+..++aaa+...++.--.-------------------
---------------------------------------------..aaaaaX.---------------------------
---------------------------------------------+o+aao+X.---------------------------
----------------------------------------------..+.+.-----------------------------
-----------------------------------------------....a----------------------------
```

```
----------------'..........+OXXXX+++++'...........................-----------
---------------'.........++++XXX+++++++'...........................----------
-------------'........++++++XOXOOOXOX++++++'.......................---------
----------'.........++++++++XOOOOOOOX++++++++++'...................'------
--------'........++X++++XXXXXXOOOXXXXXXXXXXX++++++'................----
------'........++XXOOXXOOOOOOOOOOOOXOOXX+++++++++++'.............---
------'.......+++XXOOOOOOOOOOOOOOOOOOOOOXXXX+++++++++++++'..........--
-----'.......+++XOOOOOOOOOOOOOOOOOOOOOOOOOXX++++++++++++XX++'........-
----.'.......++XOOOOOOOOOOOOOOOOOOOOOOOOOOOXXX++XXXOXXXXXX++++'........-
----'.......++XXOOOOOOOOOOOOOOOOOOOOOOOOOOOOXXXXOOOOOOOXX+++++++'.......'
----'.......+++XOOOOOOOOOOOOOOOOOOOOOOOOOOOOOOXOOOOOOOOOOXOXX++++++++'.'+'.
---'.......+++++XXOOOOOOOOOOOOOOOOOOOOOOOOOOOOOOOOOOOOOOOOOOOOOOOOOOOOO
----'.......++++XOOOOOOOOOOOOOOOOOOOOOOOOOOOOOOOOOOOOOOXOXXX++++++++'.'+'.
---'.......+++XOOOOOOOOOOOOOOOOOOOOOOOOOOOOOOXOOOOOOOOOOXOXX++++++++'.....
----'.......++XXOOOOOOOOOOOOOOOOOOOOOOOOXXXXOOOOOOOXX+++++++'..........
-----.'.......++XOOOOOOOOOOOOOOOOOOOOOOOOXXX++XXXOXXXXXX+++++'...........-
------'.......++++XOOOOOOOOOOOOOOOOOOOOOOOXX+++++++++++++XX++'...........--
-------'.......+++XXOOOOOOOOOOOOOOOOOOXXXX++++++++++++'...............---
--------'.......++XXXOOXXOOOOOOOOOOOOXOOXX+++++++++++'..............----
---------'.......++X++++XXXXXOOOXXXXXXXXXXXX++++'...................------
-----------'.........++++++++XOOOOOOOX++++++++++'.................-------
------------'.........+++++XOXOOOXOX++++++'.......................--------
--------------'........'..+++XXXX+++++++'.........................---------
```

Chapter 8

1.

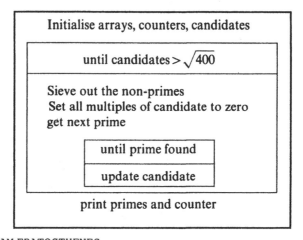

```
PROGRAM ERATOSTHENES
! all primes less than limit by the sieve of Eratosthenes
IMPLICIT NONE
INTEGER::PRIME_COUNT,I,K,PRIME_CANDIDATE,LIMIT
INTEGER,ALLOCATABLE::FIRST_N(:)
READ*,LIMIT
ALLOCATE(FIRST_N(LIMIT))
  FIRST_N=(/(I,I=1,LIMIT)/)
  PRIME_COUNT=0
  PRIME_CANDIDATE=2
! sieve out all the non-primes
    DO
      PRINT*,FIRST_N(PRIME_CANDIDATE)
      PRIME_COUNT=PRIME_COUNT+1
! set all multiples of candidate to zero
      FIRST_N(2*PRIME_CANDIDATE:LIMIT:PRIME_CANDIDATE)=0
```

```
! get the next prime
    DO
        PRIME_CANDIDATE = PRIME_CANDIDATE + 1
        IF (FIRST_N (PRIME_CANDIDATE) / = 0) EXIT
    END DO
    IF (PRIME_CANDIDATE > SQRT (LIMIT + .001)) EXIT
    END DO
! print the remaining primes
    DO K = PRIME_CANDIDATE, LIMIT
        IF (FIRST_N (K) / = 0) THEN
            PRINT*, K
            PRIME_COUNT = PRIME_COUNT + 1
        ENDIF
    END DO
    PRINT*, 'THERE ARE', PRIME_COUNT, 'PRIMES BETWEEN 0 AND', LIMIT
END PROGRAM ERATOSTHENES
```

There are 78 primes between 0 and 400.

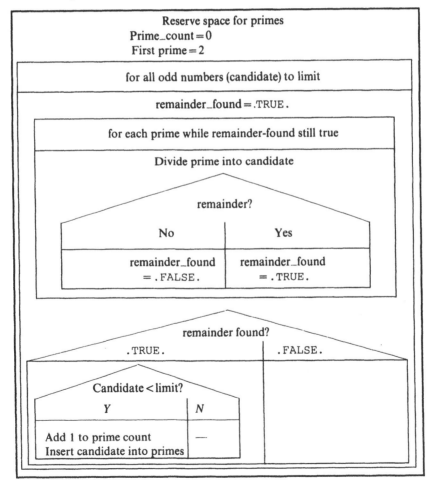

```
PROGRAM PRIMES
! yet another version of a primes program
! this one doesn't store all candidates
IMPLICIT NONE
INTEGER :: CANDIDATE, SPACE, LIMIT, PRIME_COUNT, INDEX
INTEGER, ALLOCATABLE :: FIRST_N(:)
REAL :: REAL_LIMIT
LOGICAL :: REMAINDER_FOUND
READ*, LIMIT
REAL_LIMIT = LIMIT
SPACE = REAL_LIMIT/4
PRINT*, 'SPACE ALLOCATED FOR THE PRIMES IS', SPACE
  ALLOCATE(FIRST_N(LIMIT))
  PRINT*, 'THE FIRST PRIME IS', 2
  PRIME_COUNT = 1
  FIRST_N(1) = 2
    DO CANDIDATE = 3, LIMIT-1, 2
      REMAINDER FOUND = .TRUE.
      INDEX = 0
      DO
        INDEX = INDEX + 1
        IF (CANDIDATE/FIRST_N(INDEX)*FIRST_N(INDEX) == CANDIDATE) &
                    REMAINDER_FOUND = .FALSE.
        IF (.NOT.REMAINDER_FOUND.OR.INDEX == PRIME_COUNT) EXIT
      END DO
      IF (REMAINDER_FOUND) THEN
        PRINT*, CANDIDATE
        IF (CANDIDATE <= LIMIT) THEN
          PRIME_COUNT = PRIME_COUNT + 1
          FIRST_N(PRIME_COUNT) = CANDIDATE
        ENDIF
      ENDIF
    END DO
  PRINT*, 'THERE ARE', PRIME_COUNT, 'PRIMES IN THE RANGE'
  END PROGRAM PRIMES
```

In the range 0 to 1000 there are 168 primes.

3. PROGRAM STATISTICS_2

```
! a repeat of Chapter 5, exercise 1, using an array to store the data
IMPLICIT NONE
INTEGER :: I, N
REAL :: SUM, SUMSQ, XMIN, XMAX, AMEAN, SDEV, RANGE
REAL, ALLOCATABLE :: X(:)
! decide how big the array should be
READ*, N
! now X can be made the right size
ALLOCATE(X(N))
! initialise the variables
READ*, X; SUM = X(1); SUMSQ = X(1)**2; XMIN = X(1); XMAX = X(1)
! having stored the data in X, calculate sums and minima/maxima
```

```
  DO I = 2, N
    SUM = SUM + X(I) ; SUMSQ = SUMSQ + X(I)**2
    XMIN = MIN(X(I), XMIN) ; XMAX = MAX(X(I), XMAX)
  END DO
AMEAN = SUM/N; SDEV = SQRT((SUMSQ - SUM**2/N)/(N-1))
                    RANGE = XMAX - XMIN
PRINT*, 'THE VALUES INPUT WERE'
PRINT*, (X(I), I = 1, N)
PRINT*, 'THE RESULTING STATISTICS ARE'
PRINT*, ' '
PRINT*, 'THE SUM IS', SUM
PRINT*, 'THE SUM OF THE SQUARES IS', SUMSQ
PRINT*, 'THE ARITHMETIC MEAN IS', AMEAN
PRINT*, 'THE STANDARD DEVIATION IS', SDEV
PRINT*, 'THE LARGEST AND SMALLEST NUMBERS ARE', XMAX, &
       'AND', XMIN, 'RESPECTIVELY'
PRINT*, 'THE RANGE IS', RANGE
END PROGRAM STATISTICS _
```

See Chapter 5 for the results.

4.
```
PROGRAM CURRENCY_CONVERSION
! a currency conversion from sterling or dollars
IMPLICIT NONE
INTEGER :: I, J, K, N, NO_OF_AMOUNTS, N4S
REAL :: RATE, TEMP, AMOUNT, FAMOUNT
CHARACTER(LEN = 24) :: CURRENCY
REAL, ALLOCATABLE :: DENOMINATIONS(:)
INTEGER, ALLOCATABLE :: QUANTITIES(:)
READ*, CURRENCY
READ*, N
! N is the number of denominations
ALLOCATE(DENOMINATIONS(N), QUANTITIES(N))
  READ*, (DENOMINATIONS(I), I = 1, N)
  READ*, RATE, NO_OF_AMOUNTS
  PRINT'(A, //, A, A, //, A, //, A, F7.4, //, A)' &
      '***********************************************', &
      '  CONVERSION OF STERLING TO  ', CURRENCY, &
      '***********************************************', &
      '       EXCHANGE RATE = ', RATE, &
      '***********************************************'
! bubble sort the demoninations
  DO I = 1, N-1
    DO J = 1, N-1
      IF(DENOMINATIONS(J) < DENOMINATIONS(J+1))THEN
        TEMP = DENOMINATIONS(J)
        DENOMINATIONS(J) = DENOMINATIONS(J+1)
        DENOMINATIONS(J+1) = TEMP
      ENDIF
    END DO
```

```
      END DO
! now the main part of the program
  DO I=1,NO_OF_AMOUNTS
    READ*,AMOUNT
    FAMOUNT=AMOUNT*RATE+.5*DENOMINATIONS(N)
    PRINT'(//,F6.2,A,F6.2,A,A,/)',&
        AMOUNT,' STERLING IS ',FAMOUNT,'  ',CURRENCY
! now compute the quantity of each denomination
    DO J=1,N
      QUANTITIES(J)=INT(FAMOUNT/DENOMINATIONS(J))
      FAMOUNT=FAMOUNT-QUANTITIES(J)*DENOMINATIONS(J)
    END DO
    N4S=4*INT(N/4)
    DO J=1,N4S,4
      PRINT'(/,A,4F10.2)','DENOMINATION',(DENOMINATIONS(K)',&
      & K=J,J+3)
      PRINT'(/,A,4I10)','NO REQUIRED',(QUANTITIES(K),K=J,J+3)
    END DO
    DO J=1,N-N4S,4
      PRINT'(/,A,4F10.2)','DENOMINATION   ',(DENOMINATIONS(K),&
      & K=N4S+1,N)
      PRINT'(/A,4I10)','NO REQUIRED ',(QUANTITITES(K),K=N4S+1,N
    END DO
    PRINT'(//,A)','****************************************
  END DO
END PROGRAM CURRENCY_CONVERSION
```

A typical output is shown on p. 70.

5. PROGRAM BARGE_STABILITY
```
!calculation of stability of a tilting pontoon
IMPLICIT NONE
REAL,ALLOCATABLE::DB(:),WL(:),THETA(:,:),STAB(:,:)
REAL::IMA,ML,MGD,MANG,B,D,OBD,OG,WT,WP,GAMAW,VD,DD,ANG,&
     DW,VP,PI,OGD
INTEGER::I,J,NDB,NWL
READ*,VP,WP,GAMAW,NDB,NWL
ALLOCATE(DB(NDB),WL(NWL),THETA(NDB,NWL),STAB(NDB,NWL))
READ*,DB;READ*,WL
PI=ACOS(-1.0)
! start the d/b ratio loop
DO I=1,NDB
  B=(VP/DB(I))**(1.0/3.0)
  D=DB(I)*B
  IMA=B**4/12.0
  DW=WP/(GAMAW*B*B)
  MANG=ATAN((D-DW)*2.0/B)
  OG=D/2.0
! start the load loop
  DO J=1,NWL
    WT=WP+WL(J)
```

```
ÓGD=((WP*OG)+(WL(J)*D))/WT
VD=WT/GAMAW
DD=VD/B/B
OBD=DD/2.0
ML=WL(J)*B/2.0
! calculate mgd and heel angle
    MGD=IMA/VD-(OGD-OBD)
    ANG=ML/(WT*MGD)
! check stability and heel angle
    THETA(I,J)=ANG*180.0/PI
    STAB(I,J)=MGD
    IF(MGD<0.0)STAB(I,J)=-1.0
    IF(ANG>MANG.OR.MGD<0.0)THETA(I,J)=-1.0
   END DO
  END DO
! write out some results
Print*,'IF THETA IS -1.0 PONTOON IS AWASH'
PRINT*,'IF MGD IS -1.0 PONTOON IS UNSTABLE'
PRINT'(/,17X,A,6E12.1,/)','D/B',(WL(I),I=1,NWL)
PRINT*,'              WL'
DO I=1,NDB
   PRINT'(/,A,F12.1,2X,6F12.2)','MGD',DB(I),(STAB(I,J),J=1,NWL)
   PRINT'(A,14X,6F12.1)','ANG',(THETA(I,J),J=1,NWL)
END DO
END PROGRAM BARGE_STABILITY
```

Typical results are tabulated below:

If theta is −1.0 pontoon is awash
If mgd is −1.0 pontoon is unstable

	d/b wl	0.0E+00	0.1E+06	0.2E+06	0.3E+06
mgd	0.1	28.07	26.94	25.89	24.92
ang		0.0	0.7	1.4	2.1
mgd	0.3	5.70	5.41	5.14	4.89
ang		0.0	2.5	5.0	7.6
mgd	0.5	2.08	1.91	1.75	1.60
ang		0.0	5.9	12.4	19.5
mgd	0.7	0.56	0.42	0.30	0.18
ang		0.0	23.8	−1.0	−1.0
mgd	0.9	−1.00	−1.00	−1.00	−1.00
ang		−1.0	−1.0	−1.0	−1.0

```
6. PROGRAM MARKS_ADJUSTMENT
   ! to remove unfairness and inconsistency
   IMPLICIT NONE
   REAL,ALLOCATABLE::MARKS(:,:),COLUMN(:)
```

```
INTEGER::I,STUDENTS,SUBJECTS,TAKERS
REAL::ASUM,MEAN,CONTROL_MEAN,SPREAD,CONTROL_SPREAD, &
      DIFFERENCE
READ*,STUDENTS,SUBJECTS
ALLOCATE(MARKS(STUDENTS,SUBJECTS),COLUMN(STUDENTS))
OPEN(10,FILE='MARKS.DAT')
READ(10,*)MARKS
PRINT*,COUNT(MARKS>=0)
! print out the marks as input
! negative e.g. -9 indicates not taken
DO I=1,SUBJECTS
  COLUMN=MARKS(:,I)
  TAKERS=COUNT(COLUMN>=0)
  ASUM=SUM(COLUMN,MASK=COLUMN>=0)
  PRINT*,TAKERS,ASUM
  MEAN=ASUM/TAKERS;SPREAD=SUM(ABS(COLUMN_MEAN), &
        MASK=COLUMN>=0)/TAKERS
  PRINT*,MEAN,SPREAD
  IF(I==1)THEN
    CONTROL_MEAN=MEAN;CONTROL_SPREAD=SPREAD
  ELSE
    DIFFERENCE=CONTROL_MEAN-MEAN
    WHERE(COLUMN>=0)COLUMN=COLUMN+DIFFERENCE
  END IF
  PRINT*,SUM(COLUMN,MASK=COLUMN>=0)/TAKERS
  MARKS(:,I)=COLUMN
END DO
DO I=1, STUDENTS
  PRINT'(10F6.2)',MARKS(I,:)
END DO
END PROGRAM MARKS ADJUSTMENT
```

The table of marks given below has been adjusted to a common mean and "spread

56.00	37.68	—	48.07
34.00	26.68	45.39	—
72.00	54.68	—	40.07
11.00	—	57.39	51.07
43.00	—	35.39	37.07
80.00	61.68	—	42.07
22.00	—	40.39	41.07
95.00	29.68	50.39	—
0.00	39.68	36.39	—
37.00	—	34.39	41.07
21.00	49.68	—	42.07

```
7. PROGRAM BUBBLE_SORT
! A simple sorting of numbers into descending order
IMPLICIT NONE
REAL::MAXIMUM; INTEGER::I,J,MAXIMUM_POSITION,N
```

```
REAL, ALLOCATABLE::ARRAY(:)
READ*,N
ALLOCATE(ARRAY(N))
READ*,ARRAY
DO J=1,N-1
  MAXIMUM=ARRAY(J);MAXIMUM_POSITION=J
  DO I=J+1,N
    IF (ARRAY(I)>MAXIMUM) THEN
      MAXIMUM=ARRAY(I);MAXIMUM_POSITION=I
    END IF
  END DO
  ARRAY(MAXIMUM_POSITION)=ARRAY(J)
  ARRAY(J)=MAXIMUM
END DO
PRINT*,'The array sorted into descending order is:'
PRINT*,ARRAY
END PROGRAM BUBBLE_SORT
```

Chapter 9

```
1. PROGRAM MERGE
   ! Program to read and merge student files
   IMPLICIT NONE
   CHARACTER(LEN=24)::FNAME1,FNAME2,FNAME3,SNAME1,SNAME2
   INTEGER::SNUMB1,SNUMB2,EOF1,EOF2
   REAL::AVERAGE1,AVERAGE2
   ! Get the names of the files and open them
   PRINT*, "Enter the names of the files to be merged and the name"
   PRINT*, "of the file to be produced.(Don't forget the quotes)"
   READ*,FNAME1,FNAME2,FNAME3
   OPEN (UNIT=10,FILE=FNAME1,STATUS='OLD',ACCESS='SEQUENTIAL')
   OPEN (UNIT=20,FILE=FNAME2,STATUS='OLD',ACCESS='SEQUENTIAL')
   OPEN (UNIT=30,FILE=FNAME3,STATUS='NEW',ACCESS='SEQUENTIAL')
   ! Read the first two records from each file
   READ (10,'(I5,1X,A,F4.2)',IOSTAT=EOF1) SNUMB1,SNAME1,AVERAGE1
   READ (20,'(I5,1X,A,F4.2)',IOSTAT=EOF2) SNUMB2,SNAME2,AVERAGE2
   ! If you haven't got to the end of either input file
   DO
     IF (EOF1/=0.OR.EOF2/=0) EXIT
       IF (SNUMB1<=SNUMB2) THEN
         WRITE (30,'(I5,1X,A,F4.1)')SNUMB1,SNAME1,AVERAGE1
         READ(10,'(I5,1X,A,F4.1)',IOSTAT=EOF1)SNUMB1,SNAME1,&
           AVERAGE1
       ELSE
         WRITE (30,'(I5,1X,A,F4.1)')SNUMB2,SNAME2,AVERAGE2
         READ (20,'(I5,1X,A,F4.1)',IOSTAT=EOF2)SNUMB2,SNAME2,&
           AVERAGE2
       END IF
   END DO
   ! If any more in file1 copy to file3
```

```
DO
  IF (EOF1/=0) EXIT
    WRITE (30,'(I5,1X,A,F4.1)')SNUMB1,SNAME1,AVERAGE1
    READ (10,'(I5,1X,A,F4.1)',IOSTAT=EOF1)SNUMB1,SNAME1,AVERAGE1
END DO
! If any more in file2 copy to file3
DO
  IF (EOF2/=0) EXIT
    WRITE (30,'(I5,1X,A,F4.1)')SNUMB2,SNAME2,AVERAGE2
    READ (20,'(I5,1X,A,F4.1)',IOSTAT=EOF2)SNUMB2,SNAME2,AVERAGE2
END DO
PRINT*,"File merging operation completed"
END PROGRAM MERGE
```

Chapter 10

```
1. PROGRAM STATISTICS_INT
   ! another repeat of Chapter 5, exercise 1, using array intrinsics
   IMPLICIT NONE
   INTEGER :: I, N
   REAL :: SUM_OF_X, SUMSQ, XMIN, XMAX, AMEAN, SDEV, RANGE
   REAL, ALLOCATABLE :: X (:)
   ! decide how big the array should be
   READ*, N
   ! now X can be made the right size
   ALLOCATE (X(N))
   ! initialise the variables
   READ*, (X(I), I=1,N)
   ! having stored the data in X, calculate sums and minima/maxima
     SUM_OF_X=SUM(X); SUMSQ=SUM(X**2)
     XMIN=MINVAL(X); XMAX=MAXVAL(X); RANGE=XMAX - XMIN
     AMEAN=SUM_OF_X/N; SDEV=SQRT((SUMSQ-SUM_OF_X**2/N)/&
          (N-1))
   PRINT*, 'THE VALUES INPUT WERE'
   PRINT*, (X(I), I=1,N)
   PRINT*, 'THE RESULTING STATISTICS ARE'
   PRINT*''
   PRINT*, 'THE SUM IS', SUM_OF_X
   PRINT*, 'THE SUM OF THE SQUARES IS', SUMSQ
   PRINT*, 'THE ARITHMETIC MEAN IS', AMEAN
   PRINT*, 'THE STANDARD DEVIATION IS', SDEV
   PRINT*, 'THE LARGEST AND SMALLEST NUMBERS ARE', XMAX, 'AND', &
          XMIN, 'RESPECTIVELY'
   PRINT*, 'THE RANGE IS', RANGE
   END PROGRAM STATISTICS_INT


2,3,4. PROGRAM STATISTICS_PROCEDURES
   ! the statistics example using a FUNCTION and a SUBROUTINE
   IMPLICIT NONE
   INTEGER :: I, N, NUMBER_GOOD, NUMBER_BAD
   REAL :: MEAN_OF_X, DEVIATION_OF_X, MEAN_OF_Y, DEVIATION_OF_Y, &
```

```
      COEFFICIENT
REAL, ALLOCATABLE :: X(:), Y(:)
READ*, N
ALLOCATE(X(N),Y(N))
  DO I=1,N
    READ*, X(I), Y(I)
  END DO
! calculate means and deviations
CALL STATISTICS (X, N, MEAN_OF_X, DEVIATION_OF_X)
CALL STATISTICS (Y, N, MEAN_OF_Y, DEVIATION_OF_Y)
NUMBER_GOOD=0
! filter out the bad pairs
  DO I=1, N
    IF (ABS(X(I) -MEAN_OF_X) <=3.*DEVIATION_OF_X.AND.&
      ABS Y(I) -MEAN_OF_Y) <=3.*DEVIATION_OF_Y) THEN
      NUMBER_GOOD=NUMBER_GOOD+1
      X (NUMBER_GOOD) =X(I); Y(NUMBER_GOOD) =Y(I)
    ENDIF
  END DO
CALL STATISTICS (X, NUMBER_GOOD, MEAN_OF_X, DEVIATION_OF_X)
CALL STATISTICS (Y, NUMBER_GOOD, MEAN_OF_Y, DEVIATION_OF_Y)
COEFFICIENT=CORRELATION (X, Y, NUMBER_GOOD)
NUMBER_BAD=N -NUMBER_GOOD
! print out the statistics
PRINT'(A, /, A, I5, /, A, I5, 5(/, A, F9.3), /, A)', &
    '********************************************************', &
        'NUMBER OF PAIRS OF X AND Y READ IN = ', N, &
        'NUMBER OF ''BAD'' PAIRS=                 ',NUMBER_BAD, &
        'MEAN OF ''GOOD'' X        =             ',MEAN_OF_X, &
        'MEAN OF ''GOOD'' Y        =             ',MEAN_OF_Y, &
        'STANDARD DEVIATION OF ''GOOD'' X=   ',DEVIATION_OF_X, &
        'STANDARD DEVIATION OF ''GOOD'' Y=   ',DEVIATION_OF_Y, &
        'CORRELATION COEFFICIENT OF ''GOOD''X,Y= ',COEFFICIENT, &
    '********************************************************'
  CONTAINS
  SUBROUTINE STATISTICS (ARRAY, N, MEAN, DEVIATION)
  ! this subroutine returns the mean and standard deviation
  ! of the first N entries in ARRAY
  REAL, INTENT (IN) :: ARRAY (:)
  INTEGER, INTENT (IN) :: N
  REAL, INTENT (OUT) :: MEAN, DEVIATION
  ! local variables
  REAL :: ASUM, SUMSQ
      ASUM=SUM(ARRAY)
      SUMSQ=SUM(ARRAY**2)
      MEAN=ASUM/N; DEVIATION=SQRT ( (SUMSQ-ASUM**2/N)/(N-1))
  RETURN
  END SUBROUTINE STATISTICS
  REAL FUNCTION CORRELATION (ARRAY1, ARRAY2, N)
  ! a function to calculate the correlation coefficient
```

```
    ! of the first N entries in ARRAY1 and ARRAY2
    REAL :: ARRAY1 (:), ARRAY2 (:)
    INTEGER :: N
    ! local variables
    REAL :: SUM1, SUM2, SUM12, SUMSQ1, SUMSQ2, NUMERATOR, DENOMINATOR
    SUM1 = 0; SUM2 = 0; SUM12 = 0; SUMSQ1 = 0; SUMSQ2 = 0
      SUM1 = SUM (ARRAY1); SUM2 = SUM(ARRAY2); SUMSQ1 = SUM (ARRAY1**2)
      SUMSQ2 = SUM (ARRAY2**2); SUM12 = SUM (ARRAY1*ARRAY2)
      NUMERATOR = N*SUM12 - SUM1*SUM2
    DENOMINATOR = SQRT (N*SUMSQ1 - SUM1**2)*SQRT (N*SUMSQ2 - SUM2**2)
        CORRELATION = NUMERATOR/DENOMINATOR
      RETURN
END FUNCTION CORRELATION
END PROGRAM STATISTICS_PROCEDURES
```

For the 5 pairs:

27.5	68.8
33.3	70.4
31.5	69.1
29.6	71.0
30.1	69.9

the statistics are:

```
 5
 0
30.4
69.84
 2.166
 0.907
 0.366
```

5.
```
PROGRAM DRAW_GRAPH
  !draw a simple graph using symbols
  IMPLICIT NONE
  REAL :: Q, DELTA, YMAX, YMIN, YRANGE, YSCALE
  REAL, ALLOCATABLE :: XYCOORD (:,:)
  INTEGER :: COLUMN, Y, YZERO, I, J, DATA_POINTS
  CHARACTER, PARAMETER :: BLANK = ' ', DOT = '.', CROSS = '*'
  CHARACTER :: LINE (101) = (/(' ',I=1,101)/), LINEY (101) = (/&
               ('.',I=1,101)/)
  COLUMN = 61
  ! generate data in array XYCOORD
  READ*, DATA_POINTS, DELTA
  ALLOCATE (XYCOORD(2, DATA_POINTS))
  Q = 0.0
  DO J = 1, DATA_POINTS
    XYCOORD (1, J) = Q
    XYCOORD (2, J) = (620.0*Q**1.3 - 0.4)*EXP (-130.0*Q) + 17.0*Q + 0.4
    Q = Q + DELTA
  END DO
  !find maximum and minimum Y values
  YMAX = XYCOORD (2, 1); YMIN = XYCOORD (2, 1)
```

```
DO I = 1, DATA_POINTS
   IF (XYCOORD(2, I) < YMIN) YMIN = XYCOORD(2, I)
   IF (XYCOORD (2, I) > YMAX) YMAX = XYCOORD (2, I)
END DO
! calculate scale factor
YRANGE = YMAX - MIN (0.0, YMIN)
YSCALE = (COLUMN - 1) / YRANGE
IF (YMIN < 0.0) THEN
   YZERO = ABS (YMIN) *YSCALE + 1
ELSE
   YZERO = 1
END IF
! start to print the graph
DO I = 1, DATA_POINTS
   Y = XYCOORD (2, I) *YSCALE + YZERO
   IF (ABS(XYCOORD(1, I)) < 0.001) THEN
      LINEY (Y) = CROSS
      PRINT' (1X, 2 (F6.3, 1X), 101A1)', XYCOORD(1,I), XYCOORD(2,I), LINEY
   ELSE
      LINE (YZERO) = DOT
      LINE (Y) = CROSS
      PRINT' (1X, 2 (F6.3, 1X), 101A1)', XYCOORD (1, I), XYCOORD (2, I), &
            LINE
```

```
0.000  0.000  *.......................................................
0.003  0.401  .            *
0.006  0.686  .                   *
0.009  0.850  .                       *
0.012  0.935  .                        *
0.015  0.973  .                        *
0.018  0.990  .                        *
0.021  0.997  .                        *
0.024  1.005  .                        *
0.027  1.016  .                        *
0.030  1.033  .                        *
0.033  1.056  .                       *
0.036  1.085  .                      *
0.039  1.118  .                      *
0.042  1.155  .                      *
0.045  1.196  .                       *
0.048  1.239  .                        *
0.051  1.284  .                         *
0.054  1.330  .                        *
0.057  1.378  .                        *
0.060  1.426  .                        *
0.063  1.476  .                         *
0.066  1.525  .                          *
0.069  1.575  .                          *
0.072  1.626  .                           *
0.075  1.676  .                           *
0.078  1.727  .                            *
0.081  1.778  .                             *
0.084  1.828  .                              *
0.087  1.879  .                               *
0.090  1.930  .                               *
0.093  1.981  .                                *
0.096  2.032  .                                 *
0.099  2.083  .                                  *
0.102  2.134  .                                   *
0.105  2.185  .                                    *
0.108  2.236  .                                     *
0.111  2.287  .                                      *
0.114  2.338  .                                       *
0.117  2.389  .                                        *
```

```
        LINE (Y) = BLANK
      END IF
   END DO
   END PROGRAM DRAW_GRAPH
```
The resulting 'graph' is shown on p 181.

6. `PROGRAM CUBES`

```
   ! To find solutions of x**3+y**3=a**3+b**3 by bubble sorting
   IMPLICIT NONE
   INTEGER::X,Y,ROW=0,M,I,UPPER,LOWER;INTEGER,PARAMETER::N=50
   INTEGER,ALLOCATABLE::ARRAY (:,:)
   M=N*(N+1)/2; ALLOCATE(ARRAY(M,3))
   ARRAY=0
      DO X=1,N
        DO Y=X,N
          ROW=ROW+1
          ARRAY(ROW,:) = (/X**3,Y**3,X**3+Y**3/)
        END DO
      END DO
      CALL ROW_SORT(ARRAY,3)      ! bubble sort the rows
      DO I=1, M-1
        IF (ARRAY(I,3) ==ARRAY (I+1,3)) THEN
          UPPER=HCF (ARRAY (I,1),ARRAY (I,2))
          LOWER=HCF (ARRAY(I+1,1),ARRAY (I+1,2))
          IF (HCF (UPPER,LOWER) ==1)&
          PRINT' (5I7)',&
          ARRAY (I,1),ARRAY(I,2),ARRAY(I+1,1),ARRAY(I+1,2),ARRAY(I,3
        END IF
      END DO
   CONTAINS
   SUBROUTINE ROW_SORT(A,K)
   ! bubble sort the rows of A into order according to k th elements
   INTEGER,INTENT(INOUT)::A(:,:)
   INTEGER::K,TEMPORARY,I=1,J,L,M,N
   LOGICAL::KEEP_GOING=.TRUE.
   M=UBOUND (A,1);N=UBOUND (A,2)
        DO
          IF (I==M.AND..NOT.KEEP_GOING) EXIT
            KEEP_GOING=.FALSE.
            DO L=1,M-I
              IF (A(L,K)>A(L+1,K)) THEN
                DO J=1,N
                  TEMPORARY=A(L,J)
                  A(L,J)=A(L+1,J)
                  A(L+1,J)=TEMPORARY
                END DO
                KEEP_GOING=.TRUE.
              END IF
            END DO
            I=I+1
```

```
        END DO
   END SUBROUTINE ROW_SORT
   INTEGER FUNCTION HCF (X,Y)
     INTEGER::X,Y,A,B,REMAINDER
       A=MAX(X,Y);B=MIN(X,Y)
       DO
          REMAINDER=MOD (A,B);A=B;B=REMAINDER
          IF (REMAINDER==0) EXIT
       END DO
       HCF=A
   END FUNCTION HCF
   END PROGRAM CUBES
```

Results:

1	1728	729	1000	1729
8	4096	729	3375	4104
1000	19683	6859	13824	20683
8	39304	3375	35937	39312
729	39304	4096	35937	40033
4913	59319	17576	46656	64232
1728	64000	29791	35937	65728

7. PROGRAM CONVERT_ROMAN

```
   !
   ! This program will convert read-in roman numerals into arabic
   !
     implicit none
     character :: nroman* (50)
     integer :: arabic
     read (*,'(a)') nroman
     arabic=roman_to_arabic (nroman)
     print*, 'The roman number', trim(nroman), '=', arabic
   contains
     integer function roman_to_arabic (nroman) result (narabic)
       implicit none
       character :: nroman*(*)
       character, parameter :: roman* (8) =' IVXLCDM'
       integer, parameter :: arabic (8) = (/0, 1, 5, 10, 50, 100, 500, &
       1000/)
       integer :: value, i, term
       value (i) =arabic (index(roman, nroman(i:i)))
       narabic=0
       if (verify(nroman, roman)/=0) stop '*** Invalid Roman &
       Number'
       do i=1, len(nroman)
         term=value (i)
         if (term==0) exit
         if (term<value(i+1)) term=-term
         narabic=narabic+term
       enddo
```

```
      end function roman_to_arabic
      END PROGRAM CONVERT_ROMAN
```

Typical output:

```
MCMXCIII
   The roman number MCMXCIII = 1993
```

8. PROGRAM NIM
```
  !
  ! A program to play the matchstick game of 'Nim'
  !
  ! integer :: piles (3) = (/9, 17, 12/)
    integer :: piles (3) = (/5, 43, 46/)
    do while (maxval (piles) /=0)
      call play_move ('Player A', piles)
      call play_move ('Player B', piles)
    enddo
  contains
  ! --------------------------------------------------------------
    subroutine play_move (player, piles)
      integer :: piles (3), pile, move, status
      real :: random
      character :: player* (*)
      status = ieor (ieor(piles(1), piles(2)), piles(3))
      if (status /=0) then
        do pile=1, 3 !-try all 3 piles
          move = piles (pile) - ieor (piles(pile), status)
          if (move > 0) exit
        enddo
        print *, player, ' moves', move, ' from pile', pile
      else
        pile = minval(maxloc(piles))    !/* the biggest pile */
        call random_number (random)
        move = 1 + random*piles (pile)   !/* any move */
        print *, player, ' sulks and moves', move, 'from pile', pile
      endif
      piles (pile) = piles(pile) - move
      write (*, '(t30, a, 3i5)')' so now the piles are:', piles
      if (maxval (piles) ==0) print *, player, ' wins !'
    end subroutine play_move
  ! --------------------------------------------------------------

  END PROGRAM NIM
```

Typical output:
```
  Player A sulks and moves 1 from pile 3
                               so now the piles are:    5    43    45
  Player B moves 3 from pile 2
                               so now the piles are:    5    40    45
  Player A sulks and moves 6 from pile 3
                               so now the piles are:    5    40    39
```

Player B moves 6 from pile 2	so now the piles are:	5	34	39
Player A sulks and moves 19 from pile 3	so now the piles are:	5	34	20
Player B moves 17 from pile 2	so now the piles are:	5	17	20
Player A sulks and moves 12 from pile 3	so now the piles are:	5	17	8
Player B moves 4 from pile 2	so now the piles are:	5	13	8
Player A sulks and moves 9 from pile 2	so now the piles are:	5	4	8
Player B moves 7 from pile 3	so now the piles are:	5	4	1
Player A sulks and moves 3 from pile 1	so now the piles are:	2	4	1
Player B moves 1 from pile 2	so now the piles are:	2	3	1
Player A sulks and moves 3 from pile 2	so now the piles are:	2	0	1
Player B moves 1 from pile 1	so now the piles are:	1	0	1
Player A sulks and moves 1 from pile 1	so now the piles are:	0	0	1
Player B moves 1 from pile 3	so now the piles are:	0	0	0

Player B wins!

```
9. PROGRAM TRAVELLING_SALESMAN
! The usual shortest route problem
IMPLICIT NONE
INTEGER::I,J,K,L,M,DISTANCE_TRAVELLED,MINIMUM_DISTANCE=HUGE(0),&
        NUMBER_OF_TOWNS
INTEGER,ALLOCATABLE::ROUTE(:),DISTANCE(:,:)
CHARACTER(LEN=11),ALLOCATABLE::TOWNS(:)
READ*,NUMBER_OF_TOWNS
ALLOCATE(ROUTE(NUMBER_OF_TOWNS+1),TOWNS(NUMBER_OF_TOWNS),&
        DISTANCE(NUMBER_OF_TOWNS,NUMBER_OF_TOWNS))
  TOWNS=(/'Sacramento','Berkeley',      &
                   'Palo Alto','San Jose  ','  Los Angeles'/)
  READ*,DISTANCE
    LOOP_FIVE_TIMES: &
    DO I=1,NUMBER_OF_TOWNS
      DO J=1,NUMBER_OF_TOWNS
        IF(J==I) CYCLE
        DO K=J,NUMBER_OF_TOWNS
          IF (K==J .OR. K==I) CYCLE
          DO L=K,NUMBER_OF_TOWNS
            IF (L==J .OR. L==K .OR. L==I) CYCLE
            DO M=1,NUMBER_OF_TOWNS
              IF (M==L .OR. M==K .OR. M==J .OR. M==I) CYCLE
              DISTANCE_TRAVELLED=DISTANCE(I,J)+DISTANCE(J,K)+&
                    DISTANCE(K,L)+DISTANCE(L,M)+DISTANCE(M,I)
              IF (DISTANCE_TRAVELLED<MINIMUM_DISTANCE) THEN
```

```
          MINIMUM_DISTANCE = DISTANCE_TRAVELLED
          ROUTE = (/I,J,K,L,M,I/)
        END IF
  END DO; END DO; END DO; END DO; END DO LOOP_FIVE_TIMES
  PRINT*,'Shortest route travelled is:'
    DO I = 1,NUMBER_OF_TOWNS + 1
      PRINT*, TOWNS(ROUTE(I))
    END DO
  PRINT*,'Distance travelled = ',MINIMUM_DISTANCE, 'miles
END PROGRAM TRAVELLING_SALESMAN
```

Shortest route travelled is:

Sacramento
Berkeley
San Jose
Los Angeles
Palto Alto
Sacramento
Distance travelled = 876 miles

Chapter 11

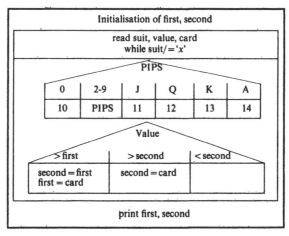

```
PROGRAM SORT_CARDS
! to sort cards into value order
IMPLICIT NONE
TYPE CARDS
    CHARACTER::SUIT
    CHARACTER::PIPS
    CHARACTER(LEN = 26)::NAME
    INTEGER::VALUE
END TYPE CARDS
INTEGER::ADD
TYPE (CARDS)::FIRST, SECOND, NEXT_CARD
FIRST = CARDS('A','1','',0)
SECOND = CARDS('A','1','',0)
```

```
      DO
        READ*, NEXT_CARD%SUIT, NEXT_CARD%PIPS, NEXT_CARD%NAME
        IF (NEXT_CARD%SUIT = = 'X') EXIT
        SELECT CASE (NEXT_CARD%SUIT)
          CASE('S');ADD=39
          CASE('H');ADD=26
          CASE('D');ADD=13
          CASE DEFAULT; ADD=0
        END SELECT
        SELECT CASE (NEXT_CARD%PIPS)
          CASE('0'); NEXT_CARD%VALUE=10
          CASE('2':'9'); NEXT_CARD%VALUE=IACHAR &
              (NEXT_CARD%PIPS) -48
          CASE('J'); NEXT_CARD%VALUE=11
          CASE('Q'); NEXT_CARD%VALUE=12
          CASE('K'); NEXT_CARD%VALUE=13
          CASE('A'); NEXT_CARD%VALUE=14
        END SELECT
        PRINT*, NEXT_CARD%VALUE
        NEXT_CARD%VALUE=NEXT_CARD%VALUE+ADD
        IF (NEXT_CARD%VALUE > FIRST%VALUE) THEN
          SECOND=FIRST
          FIRST=NEXT_CARD
         ELSE IF (NEXT_CARD%VALUE > SECOND%VALUE) THEN
          SECOND=NEXT_CARD
        ENDIF
      ENDDO
    PRINT*, FIRST%VALUE, SECOND%VALUE
    END PROGRAM SORT_CARDS


2. PROGRAM HANOI
!
! A program to solve the 'Towers of Hanoi' problem
!
  IMPLICIT NONE
  INTEGER::PIECES   ; READ*, PIECES
  CALL MOVE_PILES (PIECES, 1, 3)
  CONTAINS
    RECURSIVE SUBROUTINE MOVE_PILES (N, FROM, TO)
      IMPLICIT NONE
        INTEGER,INTENT(IN) :: N, FROM, TO
        INTEGER :: OTHER_PILE
          IF (N /= 0) THEN
            OTHER_PILE = 6 - FROM - TO
            CALL MOVE_PILES (N-1, FROM, OTHER_PILE)
            CALL MOVE_PIECE (FROM, TO)
            CALL MOVE_PILES (N-1, OTHER_PILE, TO)
          END IF
    END SUBROUTINE MOVE_PILES
    SUBROUTINE MOVE_PIECE (FROM, TO)
```

```
! ----   Try replacing this with a graphical display -----------
      IMPLICIT NONE
        INTEGER, INTENT (IN) :: FROM, TO
        INTEGER:: I = 0
          I = I + 1
          PRINT*, I, 'Move piece from  pile', FROM, 'to pile', TO
      END SUBROUTINE MOVE_PIECE
    END PROGRAM HANOI
```

Results:

```
 1) move piece from pile 1 to pile 3
 2) move piece from pile 1 to pile 2
 3) move piece from pile 3 to pile 2
 4) move piece from pile 1 to pile 3
 5) move piece from pile 2 to pile 1
 6) move piece from pile 2 to pile 3
 7) move piece from pile 1 to pile 3
 8) move piece from pile 1 to pile 2
 9) move piece from pile 3 to pile 2
10) move piece from pile 3 to pile 1
11) move piece from pile 2 to pile 1
12) move piece from pile 3 to pile 2
13) move piece from pile 1 to pile 3
14) move piece from pile 1 to pile 2
15) move piece from pile 3 to pile 2
16) move piece from pile 1 to pile 3
17) move piece from pile 2 to pile 1
18) move piece from pile 2 to pile 3
19) move piece from pile 1 to pile 3
20) move piece from pile 2 to pile 1
21) move piece from pile 3 to pile 2
22) move piece from pile 3 to pile 1
23) move piece from pile 2 to pile 1
24) move piece from pile 2 to pile 3
25) move piece from pile 1 to pile 3
26) move piece from pile 1 to pile 2
27) move piece from pile 3 to pile 2
28) move piece from pile 1 to pile 3
29) move piece from pile 2 to pile 1
30) move piece from pile 2 to pile 3
31) move piece from pile 1 to pile 3
```

3. MODULE PROCEDURES

```
CONTAINS
SUBROUTINE STATISTICS(ARRAY, N, MEAN, DEVIATION)
! This subroutine returns the mean and standard deviation
! of the first N entries in ARRAY
REAL, INTENT(IN) :: ARRAY(:)
INTEGER, INTENT(IN) :: N
REAL, INTENT(OUT) :: MEAN, DEVIATION
```

```
!local variables
REAL::ASUM,SUMSQ
        ASUM=SUM(ARRAY)
        SUMSQ=SUM(ARRAY**2)
        MEAN=ASUM/N;DEVIATION=SQRT((SUMSQ-ASUM**2/N)/(N-1))
RETURN
END SUBROUTINE STATISTICS
REAL FUNCTION CORRELATION(ARRAY1,ARRAY2,N)
! A function to calculate the correlation coefficient
! of the first N entries in ARRAY1 and ARRAY2
REAL::ARRAY1(:),ARRAY2(:)
INTEGER::N
! Local variables
REAL::SUM1,SUM2,SUM12,SUMSQ1,SUMSQ2,NUMERATOR,DENOMINATOR
SUM1=0;SUM2=0;SUM12=0;SUMSQ1=0;SUMSQ2=0
        SUM1=SUM(ARRAY1);SUM2=SUM(ARRAY2);SUMSQ1=SUM(ARRAY1**2)
        SUMSQ2=SUM(ARRAY2**2);SUM12=SUM(ARRAY1*ARRAY2)
        NUMERATOR=N*SUM12-SUM1*SUM2
   DENOMINATOR=SQRT(N*SUMSQ1-SUM1**2)*SQRT(N*SUMSQ2-SUM2**2)
   CORRELATION=NUMERATOR/DENOMINATOR
 RETURN
END FUNCTION CORRELATION
END MODULE PROCEDURES
PROGRAM STATISTICS_PROCEDURES_2
! The statistics example using a FUNCTION and a SUBROUTINE
! This version uses a MODULE
USE PROCEDURES
IMPLICIT NONE
INTEGER ::I,N,NUMBER_GOOD,NUMBER_BAD
REAL::MEAN_OF_X,DEVIATION_OF_X,MEAN_OF_Y,DEVIATION_OF_Y,&
        COEFFICIENT
REAL,ALLOCATABLE::X(:),Y(:)
READ*,N
ALLOCATE(X(N),Y(N))
  DO I=1,N
    READ*,X(I),Y(I)
  END DO
! Calculate means and deviations
CALL STATISTICS(X,N,MEAN_OF_X,DEVIATION_OF_X)
CALL STATISTICS(Y,N,MEAN_OF_Y,DEVIATION_OF_Y)
NUMBER_GOOD=0
! Filter out the bad pairs
    DO I=1,N
      IF (ABS(X(I)-MEAN_OF_X)<=3.*DEVIATION_OF_X.AND.&
          ABS(Y(I)-MEAN_OF_Y)<=3.*DEVIATION_OF_Y) THEN
          NUMBER_GOOD=NUMBER_GOOD+1
          X(NUMBER_GOOD)=X(I);Y(NUMBER_GOOD)=Y(I)
      ENDIF
    END DO
CALL STATISTICS(X,NUMBER_GOOD,MEAN_OF_X,DEVIATION_OF_X)
CALL STATISTICS(Y,NUMBER_GOOD,MEAN_OF_Y,DEVIATION_OF_Y)
COEFFICIENT=CORRELATION(X,Y,NUMBER_GOOD)
NUMBER_BAD=N-NUMBER_GOOD
! Print out the statistics
PRINT'(A,/,A,I5,/,A,I5,5(/,A,F9.3),/,A)',&
```

```
'*****************************************************',&
'Number of pairs of x and y read in = ',N,&
'Number of "bad" pairs             = ',NUMBER_BAD,&
'Mean of "good" x                  = ',MEAN_OF_X,&
'Mean of "good" y                  = ',MEAN_OF_Y,&
'Standard Deviation of "good" x = ',DEVIATION_OF_X,&
'Standard Deviation of "good" y  = ',DEVIATION_OF_Y,&
'Correlation Coefficient of "good" x,y = ',COEFFICIENT,&
'*****************************************************',
```

END PROGRAM STATISTICS_PROCEDURES_2

Chapter 13

```
1. PROGRAM NEWTON_RAPHSON
   !solves a single nonlinear equation by the 'Newton-Raphson' method
   IMPLICIT NONE
   REAL::X0,X1,TOL
   INTEGER::ITERS,ITS
   LOGICAL::CONVERGED
   READ*,X0,TOL,ITS
   PRINT*,'*************NEWTON-RAPHSON METHOD******************'
   PRINT*,'INITIAL VALUE'
   PRINT*,X0
   ITERS=0
   DO
     ITERS=ITERS+1
     X1=X0-F(X0)/FDASH(X0)
     CONVERGED=(ABS(X1-X0)/ABS(X1)<TOL)
     X0=X1
     IF(CONVERGED.OR.ITERS==ITS)EXIT
   END DO
   PRINT*,'SOLUTION AND ITERATIONS TO CONVERGENCE'
   PRINT*,X1,'   ',ITERS
   CONTAINS
   REAL FUNCTION F(X)
   ! this varies with the problem being solved
   REAL::X
   F=(620.0*X**1.3-0.4)*EXP(-130.0*X)+17.0*X-1.1
   RETURN
   END FUNCTION F
   REAL FUNCTION FDASH(X)
   ! the derivative also varies of course
   REAL::X
   FDASH=EXP(-130.0*X)*(-620.0*130.0*X**1.3+620.0*1.3*X**0.3 &
   +.4*130.)+17.0
   RETURN
   END FUNCTION FDASH
   END PROGRAM NEWTON_RAPHSON
```

The strain for a stress of 1.5Pa is 0.064475

2.
```
PROGRAM BISECTION
! solves a single nonlinear equation by the 'bisection' method
IMPLICIT NONE
REAL::X0,X1,XMID,XOLD,TOL;INTEGER::ITERS,ITS;LOGICAL::CONVERGED
READ*,X0,X1,TOL,ITS
PRINT*,'*******************BISECTION METHOD*******************'
PRINT*,'INITIAL VALUES'
PRINT*,X0,'   ',X1
ITERS=0;XOLD=X0
DO
  ITERS=ITERS+1
  XMID=0.5*(X0+X1)
  IF(F(X0)*F(XMID)<.0) THEN
    X1=XMID
  ELSE
    X0=XMID
  ENDIF
  CONVERGED=(ABS(XMID-XOLD)/ABS(XMID)<TOL)
  XOLD=XMID
  IF(CONVERGED.OR.ITERS==ITS) EXIT
END DO
PRINT*,'SOLUTION AND ITERATIONS TO CONVERGENCE'
PRINT*.X1,'   ',ITERS
CONTAINS
REAL FUNCTION F(X)
! this varies with the problem being solved
REAL::X
F=(620.0*X**1.3-0.4)*EXP(-130.0*X)+17.0*X-1.1
RETURN
END FUNCTION F
END PROGRAM BISECTION
```

The strain found is the same as in case study 1.

3,4
```
PROGRAM GAUSSIAN_ELIMINATION
--------------L-U factorisation of an n×n system---------------
    IMPLICIT NONE
    REAL,ALLOCATABLE::A(:,:),X(:),B(:)
    INTEGER::N
    READ*,N    ; ALLOCATE(A(N,N),X(N),B(N))
    READ*,A; READ*,B
    CALL ELIMINATE(A,X,B)  ; PRINT*,X
  CONTAINS
!---------------------------------------------------------------
    SUBROUTINE ELIMINATE(A,X,B)
    REAL,INTENT(INOUT) ::A(:,:)  ! The stiffness matrix
    REAL,INTENT(IN)    ::B(:)    ! The unknown LHS
    REAL,INTENT(OUT)   ::X(:)    ! The given RHS
```

```
      INTEGER::I,J,N;   REAL::ASUM
      N=UBOUND(A,1)
!-----------------------------------------------------------------
!----- factorisation stage
!----- l and u are stored in a; it is assumed that l(i,i)=1.0
!----- and so it is not stored
      DO I=2,N
        DO J=I,N
          ASUM=SUM( A(J,1:I-2)*A(1:I-2,I-1))
          A(J,I-1)=(A(J,I-1)-ASUM)/A(I-1,I-1)
          A(I,J)=A(I,J)-SUM(A(I,1:I-1)*A(1:I-1,J))
        END DO
      END DO
!-----------------------------------------------------------------
      DO I=1,N                        ! forward substitution stage
        X(I)=B(I)                     ! this depends on the
        B(I+1:N)=B(I+1:N)&            ! assumption : L(I,I)=1.0
                -A(I+1:N,I)*X(I)
      ENDDO
!-----------------------------------------------------------------
      DO I=N,1,-1                     ! backward substitution stage
        X(I)=B(I)/A(I,I)
        B(N-1:1:-1)=B(N-1:1:-1)-A(N-1:1:-1,I)*X(I)
      END DO
!-----------------------------------------------------------------
    END SUBROUTINE ELIMINATE
  END PROGRAM GAUSSIAN_ELIMINATION
```

The vertical deflection of node 6 (x_{10}) is 40.4444 units.

Due to the symmetry of A and the number of zero coefficients it possesses, "band storage strategies, which take account of symmetry, should be used.

```
5. PROGRAM CONJUGATE_GRADIENTS
   ! a conjugate gradient algorithm
   IMPLICIT NONE
   INTEGER::ITERS,ITS,N;LOGICAL::CONVERGED
   REAL::TOL,UP,ALPHA,BETA
   REAL,ALLOCATABLE::A(:,:),B(:),X(:),R(:),U(:),P(:),XNEW(:)
   READ*,N,TOL,ITS
   ALLOCATE(A(N,N),B(N),X(N),R(N),U(N),P(N),XNEW(N))
   OPEN(10,FILE='TRUSS.DAT')
   READ(10,*)A; READ(10,*)B
   X=1.0
   R=B-MATMUL(A,X);P=R
   ITERS=0
   DO
     ITERS=ITERS+1
     U=MATMUL(A,P);UP=DOT_PRODUCT(R,R)
```

```
      ALPHA=UP/DOT_PRODUCT(P,U)
      XNEW=X+P*ALPHA
      R=R-U*ALPHA;BETA=DOT_PRODUCT(R,R)/UP
      P=R+P*BETA
      CONVERGED=(MAXVAL(ABS(XNEW-X))/MAXVAL(ABS(XNEW))<TOL)
      X=XNEW
      IF(CONVERGED.OR.ITERS==ITS)EXIT
    END DO
  PRINT*,ITERS
  PRINT*,X
END PROGRAM CONJUGATE_GRADIENTS
```

For TOL in the above program equal to 10^{-5}, the solution $x_{10} = 40.4444$ is obtained in 16 iterations. As in Exercise 6 there is scope for only storing the non-zero entries in A, for considering symmetry when carrying out the step $u = Ap^k$ and even not assembling A at all in this and similar algorithms.

```
6. PROGRAM COMPLEX_GAUSSIAN_ELIMINATION
   ! L-U factorisation of a complex n×n system
   IMPLICIT NONE
   REAL,ALLOCATABLE::RA(:,:),IA(:,:),RB(:),IB(:)
   COMPLEX,ALLOCATABLE::A(:,:),X(:),B(:)
   INTEGER::N
   READ*,N
   ALLOCATE(RA(N,N),IA(N,N),A(N,N),X(N),RB(N),IB(N),B(N))
     READ*,RA;READ*,IA
     A=CMPLX(RA,IA)
     READ*,RB;READ*,IB
     B=CMPLX(RB,IB)
   CALL ELIMINATE(A,X,B)
   PRINT*,X
   CONTAINS
   SUBROUTINE ELIMINATE(A,X,B)
   COMPLEX,INTENT(IN)::B(:);COMPLEX,INTENT(OUT)::X(:)
   COMPLEX,INTENT(INOUT)::A(:,:)
   INTEGER::I,J,K,N;COMPLEX::SUM
   N=UBOUND(A,1)
   ! factorisation stage
   ! l and u are stored in a; it is assumed that l(i,i)=1.0
   ! and so it is not stored
     DO I=2,N
       DO J=I,N
         SUM=0.0
         DO K=1,I-2
         SUM=SUM+A(J,K)*A(K,I-1)
       END DO
       A(J,I-1)=(A(J,I-1)-SUM)/A(I-1,I-1)
       SUM=0.0
       DO K=1,I-1
         SUM=SUM+A(I,K)*A(K,J)
```

```
   END DO
   A(I,J)=A(I,J)-SUM
 END DO
END DO
! forward substitution stage
! this depends on the assumption l(i,i)=1.0
 DO I=1,N
   X(I)=B(I)
   B(I+1:N)=B(I+1:N)-A(I+1:N,I)*X(I)
 END DO
! backward substitution stage
 DO I=N,1,-1
   X(I)=B(I)/A(I,I)
   B(N-1:1:-1)=B(N-1:1:-1)-A(N-1:1:-1,I)*X(I)
 END DO
END SUBROUTINE ELIMINATE
END PROGRAM COMPLEX_GAUSSIAN_ELIMINATION
```

The solution is:

$$
\begin{bmatrix}
(\frac{3}{2}, & -\frac{1}{2}) \\
(1.0, & 1.0) \\
(-\frac{1}{2}, & -\frac{5}{2})
\end{bmatrix}
$$

7.
```
PROGRAM EIGENVALUES
! compute natural frequencies by a simple iterative method
IMPLICIT NONE
REAL,ALLOCATABLE::A(:,:),X0(:),X1(:)
INTEGER::,N,ITERS,ITS;LOGICAL::CONVERGED
READ::TOL,BIG
READ*,N,TOL,ITS
ALLOCATE(A(N,N),X0(N),X1(N))
READ*,A;READ*,X0
ITERS=0
  DO
    ITERS=ITERS+1
    X1=MATMUL(A,X0)
    BIG=MAXVAL(X1);IF(ABS(MINVAL(X1))>BIG)BIG=MINVAL(X1
        X1=X1/BIG
    CONVERGED=(MAXVAL(ABS(X1-X0))/MAXVAL(ABS(X1))<TOL)
    X0=X1
    IF(CONVERGED.OR.ITERS==ITS)EXIT
  END DO
X1=X1/SQRT(SUM(X1**2))
PRINT*,BIG;PRINT*,X1;PRINT*,ITERS
END PROGRAM EIGENVALUES
```

For a "tolerance" TOL of 10^{-6}, λ converges to 6.4673 in 8 iterations.

8.
```
PROGRAM RUNGE_KUTTA
! 4th order method for systems of equations
IMPLICIT NONE
REAL,ALLOCATABLE::Y(:),Y0(:),K0(:),K1(:),K2(:),K3(:)
INTEGER::N,STEPS,I,J
REAL::H,X
READ*,N,STEPS,H,X
ALLOCATE(Y(N),Y0(N),K0(N),K1(N),K2(N),K3(N))
READ*,Y
PRINT*,'*****************SYSTEMS OF EQUATIONS*****************'
PRINT*,'**********4TH ORDER RUNGE-KUTTA METHOD**********'
PRINT*,'  X  Y(I)   ,I=1,',N
DO J=0,STEPS
   PRINT'(5E13.5)',X,(Y(I),I=1,N)
   K0=FUNC(X,Y,N);Y0=Y;Y=Y0+.5*H*K0;X=X+.5*H
   K1=FUNC(X,Y,N);Y=Y0+.5*H*K1
   K2=FUNC(X,Y,N);Y=Y0+H*K2;X=X+.5*H
   K3=FUNC(X,Y,N);Y=Y0+(K0+2.*(K1+K2)+K3)/6.*H
END DO
CONTAINS
FUNCTION FUNC(X,Y,N)
! provides the values of f(x,y(i)) specified by the user
IMPLICIT NONE
INTEGER,INTENT(IN)::N
REAL::FUNC(N)
REAL,INTENT(IN)::X,Y(:)
FUNC(1)=3.0*X*Y(2)+4.0
FUNC(2)=X*Y(1)-Y(2)-EXP(X)
RETURN
END FUNCTION FUNC
END PROGRAM RUNGE_KUTTA
```

The results are $y(0.5) = 6.2494$, $z(0.5) = 0.67386$.

9.
```
PROGRAM SURGE_TANK
! oscillations in a tank by a 4th order method for systems of
equations
IMPLICIT NONE
REAL,ALLOCATABLE::Y(:),Y0(:),K0(:),K1(:),K2(:),K3(:)
INTEGER::N,STEPS,I,J
REAL::H,X
READ*,N,STEPS,H,X
ALLOCATE(Y(N),Y0(N),K0(N),K1(N),K2(N),K3(N))
READ*,Y
PRINT*,'*****************SYSTEMS OF EQUATIONS*****************'
PRINT*,'**********4TH ORDER RUNGE-KUTTA METHOD**********'
PRINT*,'  x  Y(I)   ,I=1,',N
DO J=0,STEPS
   PRINT'(5E13.5)',X,(Y(I),I=1,N)
   K0=FUNC(X,Y,N);Y0=Y;Y=Y0+.5*H*K0;X=X+.5*H
   K1=FUNC(X,Y,N);Y=Y0+.5*H*K1
```

```
      K2=FUNC(X,Y,N);Y=Y0+H*K2;X=X+.5*H
      K3=FUNC(X,Y,N);Y=Y0+(K0+2.*(K1+K2)+K3)/6.*H
END DO
CONTAINS
FUNCTION FUNC(X,Y,N)
! provides the values of f(x,y(i)) specified by the user
IMPLICIT NONE
INTEGER,INTENT(IN)::N
REAL::FUNC(N)
REAL,INTENT(IN)::X,Y,(:)
! local variables and constants
REAL,PARAMETER::G=9.81
REAL::BIG_D,SMALL_D,F,L,BIG_A,SMALL_A,PI
BIG_D=2.5;SMALL_D=0.8;L=1200.0;F=0.005;PI=4.*ATAN(1.)
BIG_A=PI*BIG_D**2/4.0;SMALL_A=PI*SMALL_D**2/4.0
FUNC(1)=-Y(2)*SMALL_A/BIG_A
FUNC(2)=(Y(1)-4*F*L*Y(2)**2/2./SMALL_D)*G/L
RETURN
END FUNCTION FUNC
END PROGRAM SURGE_TANK
```

The initial conditions for case 1 might be $y(0) = 150.0$, $v(0) = 10.0$. Under these circumstances, the water level in the surge tank reaches 7 m above reservoir level in about 350 seconds after turbine shut-down. In the second case the initial conditions are $y(0) = 0$, $v(0) = 0$ and the water level in the tank drops to 70 m below reservoir level about 100 seconds after start-up.

10. The Structure Chart is as given in the text

```
PROGRAM DATE_OF_EASTER
! a program to calculate the date of easter
  INTEGER::NUMBER_OF_YEARS,YEAR_WANTED,DAY
  CHARACTER::*8 MONTH,DESIGNATION
  READ*,NUMBER_OF_YEARS
    DO I=1,NUMBER_OF_YEARS
      READ*,YEAR_WANTED
      CALL EASTER(YEAR_WANTED,MONTH,DAY)
        SELECT CASE (DAY)
        CASE (1,21,31)
          DESIGNATION='ST'
        CASE(2,22)
          DESIGNATION='ND'
        CASE(3,23)
          DESIGNATION='RD'
        CASE DEFAULT
          DESIGNATION='TH'
        END SELECT
      PRINT'(A,I5,A,I3,A,A)','EASTER DAY IN THE YEAR',YEAR_WANTED&
        ,' FALLS ON',DAY,DESIGNATION,MONTH
    END DO
  CONTAINS
```

```
SUBROUTINE EASTER(YEAR,MONTH,DAY)
easter day in any given year is the sunday following the
paschal moon, which is the full moon on or after march 21st.
to calculate the day of the week of the paschal moon, use the
''dominical number'', which is the date of the first sunday in january.
a full moon is said to occur 14 days after the preceding new moon.
the age of the moon on january 1st. (i.e. the number of days since
the last new moon is called the ''epact'', and from it the dates
of the new moons in each month may be obtained.
the pattern of epacts repeats every 19 years, the ''golden number''
being the position within the cycle. lilius computed a correction
because the cycle is really 19 years less 1.5 hours.
    IMPLICIT NONE
    INTEGER,INTENT(IN) ::YEAR;INTEGER,INTENT(OUT) ::DAY
    CHARACTER,INTENT(OUT) ::MONTH*8
      INTEGER::LEAP,CENTURIES,DOMINICAL_NUMBER,GOLDEN_NUMBER,&
        LILIAN_CORRECTION,CLAVIAN_CORRECTION,JANUARY_MOON,&
        MARCH_MOON,APRIL_MOON,MARCH_21ST,MARCH_START,EPACT,&
        APRIL_START,PASCHAL_MOON,PASCHAL_DAY,EASTER_DAY,PLUS
      IF(MOD(YEAR,4)==0.AND.MOD(YEAR,100)/=0&
        .OR.MOD(YEAR,400)==0)THEN
          LEAP=1
        ELSE
          LEAP=0
      ENDIF
      CENTURIES=YEAR/100
      DOMINICAL_NUMBER=7-MOD(YEAR+YEAR/4-CENTURIES&
        +CENTURIES/4-1-LEAP,7)
      GOLDEN_NUMBER=MOD(YEAR,19)+1
      LILIAN_CORRECTION=MOD(CENTURIES-CENTURIES/4&
        -(CENTURIES-(CENTURIES-17)/25)/3-8,30)
      EPACT=MOD((GOLDEN_NUMBER-1)*11-LILIAN_CORRECTION,30)
        IF(GOLDEN_NUMBER>11)THEN
          CLAVIAN_CORRECTION=26
        ELSE
          CLAVIAN_CORRECTION=25
        ENDIF
dates are held as number of days from the beginning of the year
      JANUARY_MOON=31-EPACT
      MARCH_MOON=JANUARY_MOON+59+LEAP
        IF(EPACT>CLAVIAN_CORRECTION)THEN
          PLUS=30
        ELSE
          PLUS=29
        ENDIF
      APRIL_MOON=MARCH_MOON+PLUS
      MARCH_21ST=31+28+LEAP+21
        IF(MARCH_MOON+13<MARCH_21ST)THEN
          PASCHAL_MOON=APRIL_MOON+13
        ELSE
```

```
              PASCHAL_MOON=MARCH_MOON+13
          ENDIF
          PASCHAL_DAY=MOD(PASCHAL_MOON-DOMINICAL_NUMBER,7)+1
          EASTER_DAY=PASCHAL_MOON+8-PASCHAL_DAY
        MARCH_START=31+28+LEAP
        APRIL_START=MARCH_START+31
          IF(EASTER_DAY<=APRIL_START)THEN
              MONTH='MARCH'
              DAY=EASTER_DAY-MARCH_START
          ELSE
              MONTH='APRIL'
              DAY=EASTER_DAY-APRIL_START
          ENDIF
          RETURN
      END SUBROUTINE EASTER
    END PROGRAM DATE_OF_EASTER
```

Easter in the year 7086 falls on 18 April.

```
11. MODULE LEAP
      CONTAINS
    LOGICAL FUNCTION LEAP_YEAR (YEAR)
    IMPLICIT NONE
    INTEGER::YEAR
    LOGICAL::YEAR_4,YEAR_100,YEAR_400
      YEAR_4=MOD(YEAR,4)==0;YEAR_100=MOD(YEAR,100)==0
      YEAR_400=MOD(YEAR,400)==0
      LEAP_YEAR=(YEAR_4.AND..NOT.YEAR_100).OR.YEAR_400
    END FUNCTION LEAP_YEAR
    END MODULE LEAP
    PROGRAM CALENDAR
    ! Outputs the calendar for a given year
    USE LEAP
    IMPLICIT NONE
    INTEGER::DAY,MONTH(12),YEAR,MON,QUARTER,R,ARRAY(7,18),I,J
    CHARACTER::DAYNAME*3,HEADING(4)*90
    MONTH=(/31,28,31,30,31,30,31,31,30,31,30,31/)
    HEADING(1)='(13X,"January",19X,"February",19X,"March")
    HEADING(2)='(14X,"April",24X,"May",25X,"June")'
    HEADING(3)='(14X,"July",24X,"August",21X,"September")
    HEADING(4)='(13X,"October",18X,"November",18X,"December")
    READ*,YEAR
    OPEN(10,FILE='CAL.OUT')
    WRITE(10,'("Calendar for",I4//)')YEAR
    IF (LEAP_YEAR(YEAR))MONTH(2)=29
    CALL ZELLER(1,1,YEAR,R,DAYNAME)
    IF (R==0)R=R+7
    MON=1;DAY=1;I=R
    DO QUARTER=1,4
      WRITE(10,HEADING(QUARTER))
      ARRAY=0
```

```
    J=1
    DO
      IF (I<=7.AND.DAY<=MONTH(MON)) THEN
        ARRAY(I,J)=DAY
        I=I+1;DAY=DAY+1
      END IF
      IF (I>7) THEN
        I=1;J=J+1
      END IF
      IF (DAY>MONTH(MON))THEN
      MON=MON+1;DAY=1
        SELECT CASE(J)
          CASE(:6)
            J=7
          CASE(7:12)
            J=13
          CASE DEFAULT
            J=19;CALL PRINTOUT(ARRAY)
        END SELECT
      END IF
      IF (J==19) EXIT
    END DO
  END DO
CONTAINS
SUBROUTINE PRINTOUT(ARRAY)
IMPLICIT NONE
INTEGER,INTENT(OUT)::ARRAY(:,:)
CHARACTER::NAME(7)*3
INTEGER::I
NAME=(/'Sun','Mon','Tue','Wed','Thu','Fri','Sat'/)
DO I=1,7
  WRITE(10,'(A3,1X,6(I3.0,1X),2X,6(I3.0,1X),2X,6(I3.0,1X))')&
          NAME(I),(ARRAY(I,J),J=1,18)
END DO
WRITE(10,'()')
WRITE(10,'()')
END SUBROUTINE PRINTOUT
SUBROUTINE ZELLER(DAY,AMONTH,YEAR,R,DAYNAME)
IMPLICIT NONE
INTEGER::DAY,YEAR,R,NY,NM,AMONTH
CHARACTER*3::DAYNAME,NAME(0:6)
NAME=(/'Sat','Sun','Mon','Tue','Wed','Thu','Fri'/)
CALL CHECK(DAY,AMONTH,YEAR)
NY=YEAR;NM=AMONTH
IF (AMONTH==1.OR.AMONTH==2) THEN
  NM=NM+12;NY=NY-1
END IF
R=MOD(DAY+2*NM+2+(3*NM+3)/5+NY+NY/4-NY/100+NY/400,7)
DAYNAME=NAME(R)
END SUBROUTINE ZELLER
```

```
SUBROUTINE CHECK(DAY,MON,YEAR)
USE LEAP
IMPLICIT NONE
INTEGER::DAY,YEAR,MONTH(12),MON
INTEGER,PARAMETER::LOW_YEAR=1583,HIGH_YEAR=10000
     MONTH=(/31,0,31,30,31,30,31,31,30,31,30,31/)
     IF (YEAR<LOW_YEAR.OR.YEAR>HIGH_YEAR) CALL ERROR(1)
     IF (LEAP_YEAR(YEAR)) THEN
       MONTH(2)=29
     ELSE
       MONTH(2)=28
     END IF
   IF (MON<1.OR.MON>12) CALL ERROR(3)
   IF (DAY<1.OR.DAY>MONTH(MON)) CALL ERROR(4)
END SUBROUTINE CHECK
SUBROUTINE ERROR(I)
IMPLICIT NONE
  INTEGER::I;INTEGER,PARAMETER::SUB=4
  CHARACTER::TEXT(SUB+1)*80
  TEXT(1)='Invalid Year';TEXT(2)='Invalid Julian'
  TEXT(3)='Invalid Month';TEXT(4)='Invalid Day'
  TEXT(5)='Subroutine Error(i) Called with invalid i'
    IF (I>=1.AND.I<=SUB) THEN
      WRITE(10,'(A)')TEXT(I)
      ELSE
      WRITE(10,'(A)')TEXT(SUB+1)
    END IF
END SUBROUTINE ERROR
END PROGRAM CALENDAR
```

The calendar for the first three months of 1995 is given below.

	January							February						March				
Sun	1	8	15	22	29	0	0	5	12	19	26	0	0	5	12	19	26	
Mon	2	9	16	23	30	0	0	6	13	20	27	0	0	6	13	20	27	
Tue	3	10	17	24	31	0	0	7	14	21	28	0	0	7	14	21	28	
Wed	4	11	18	25	0	0	1	8	15	22	0	0	1	8	15	22	29	
Thu	5	12	19	26	0	0	2	9	16	23	0	0	2	9	16	23	30	
Fri	6	13	20	27	0	0	3	10	17	24	0	0	3	10	17	24	31	
Sat	7	14	21	28	0	0	4	11	18	25	0	0	4	11	18	25	0	

12. PROGRAM REVERSE_POLISH_NOTATION

```
character::expr*(80),rpn*(80),stack(80),symbol
integer::top
! integer priority
do
print*,'Enter an expression:'; read(*,'(a)') expr
if expr.eq.'    ') exit
ipos=1
top=1 rpn='  '
```

```fortran
do i = 1, len(trim(expr))
  symbol = expr(i:i)
  select case (symbol)
  case(' ');                          !/*skip spaces*/
case('(')
  call push (symbol,stack,top)
  case (')')
    do                               ! /*pop stack symbols  */
      if (stack(top) == '(') exit  !/*encountered        */
      symbol = pop (stack,top)       !/*until a ')' is      */
      call write_symbol (symbol,rpn,ipos)
    enddo
    symbol = pop (stack,top)   !/* discard ')' too */
  case ('A':'Z','a':'z','0':'9','.')
    call write_symbol (symbol,rpn,ipos)
  case ('+','-','*','/','^')
    do
      if (top == 0 .or. priority(symbol) >= priority (stack(top)))&
                exit
      call write_symbol (pop(stack,top),rpn,ipos)
    enddo
    call push(symbol,stack,top)
  case default
    print*, 'invalid character found: (' //symbol//')'
    stop
  end select
enddo
do
  if (top == 0) exit
  call write_symbol (pop(stack,top),rpn,ipos)
enddo
print*, rpn
enddo
contains
!-------------------------------------------------------------
  subroutine write_symbol(symbol,rpn,ipos)
    character :: symbol, rpn*(*)
    integer :: ipos
    rpn (ipos:ipos+1) = ' '//symbol
    ipos = ipos+2
  end subroutine write_symbol
!-------------------------------------------------------------
  integer function priority (oper)
    character :: oper
    select case (oper)
    case ('(')      ;    priority = 0
    case ('+','-'); priority = 1
    case ('*','/')  ;   priority = 2
    case ('^')     ;    priority = 3
    case default   :    priority = -1
```

```
    end select
    end function priority
!-----------------------------------------------------------------
    subroutine push (symbol,stack,top)
      character : : symbol,stack(*)
      integer : : top
      top=top+1; stack(top)=symbol
    end subroutine push
!-----------------------------------------------------------------
    character function pop (stack,top) result (symbol)
      character : : stack(*)
      integer : : top
      symbol=stack(top); top=top-1
    end function pop
!-----------------------------------------------------------------
!-----------------------------------------------------------------
END PROGRAM REVERSE_POLISH_NOTATION
```

The expression $x + y*z$ is rendered as $xyz*+$ and so on

13. PROGRAM RURITANIA_STAMPS

```
    !
    ! This solves the 'Ruritania Postage Stamps' problem
    !
    integer, parameter : : n=7 !/* number of different stamps*/
    integer : : set(0:n), best_set(0:n), best_total,ic
    logical : : finished=.false.
    character : : fmt*25
    write (fmt,'(a,i2,a)')'(i4,a,',n,'i4,a,i5,a,i5)'
    print*,fmt
    call setup (set,n)
    best_total=0
    do ic=1,huge(ic)
      max_value=Get_Max_Val (set,n)
      if (max_value>best_total) then
        best set=set
        best_total=max_value
      endif
!  write(*,fmt) ic,'set=',set(1:n),'max val=',Max_value &
!  &  ,'best=',best_total
      call Generate_Next_Set (set,n, max_value,best_total,finished)
      if (finished) exit
    enddo

    print*,ic,'combinations were tried:'
    print*,'The best set of stamps was', best_set
    print*,'With a maximum value of:',best_total
contains
!-----------------------------------------------------------------
    Integer Function Get_Max_Val (set,n) result (max_val)
```

```
integer::set(0:n)
logical::postage(0:3*set(n)+1)
max_val=-1
postage=.false.
do i=0,n
  do j=i,n
    do k=j,n
      postage(set(i)+set(j)+set(k))=.true.
    enddo
  enddo
enddo
do i=1,3*set(n)+1
  if (.not.postage(i)) then
    max_val=i-1
    return
  endif
enddo
End Function Get_MAx_Val
```

```
Subroutine Generate_Next_Set (set,n, max_value,best_total,
finished)

This will produce the next set of stamps to try
with 'finished' set to .true. when no more sets are possible
  logical::finished
  integer::set(0:n), n, best_total
if (set(n)>max_value) then ! -speed up the search
  do i=n,1,-1
  enddo if (set(i-1)<max_value) exit
i=i+2    !/*fudge up i->i+1*/
3 set(i)=max_value+1
  do j=i+1,n
    set(j)=3*set(j-1)+1    !/*reset values to the right*/
  enddo

else !------------------------------------------------------
2 do i=n,2,-1
  if (set(i)/=set(i-1)+1) goto 1  !/*room to decrement?*/
  enddo
  finished=.true.
  return
1 set(i)=set(i)-1      !/*decrement*/
    do j=i+1,n     !/* and */
      set(j)=3*set(j-1)+1   !/*reset values to the right*/
    enddo
  if (3*set(n)<=best_total) then  !
    set(n)=set(n-1)+1
    goto 2
  endif
```

```
  endif    ! -2 methods of modifying the set
  End Subroutine Generate_Next_Set
!-------------------------------------------------------------
  Subroutine Setup (set,n)
!
!   gives the initial value for the set of stamps
!
    integer :: set(0:n)
    set(0)=0
    set(1)=1
    do i=2,n
      set(i)=3*set(i-1)+1
    enddo
  End Subroutine setup
!-------------------------------------------------------------
END PROGRAM RURITANIA_STAMPS
!-------------------------------------------------------------
```

When there are 5 stamps, the best set is 1, 4, 6, 14, 15 with a maximum value of 36.

```
  n=2
  2 combinations were tried:
    The best set of stamps was 0 1 3
    With a maximum value of: 7
  n=3
  10 combinations were tried:
    The best set of stamps was 0 1 4 5
    With a maximum value of: 15
  n=4
  68 combinations were tried:
    The best set of stamps was 0 1 4 7 8
    With a maximum value of: 24
  n=5
  619 combinations were tried:
    The best set of stamps was 0 1 4 6 14 15
    With a maximum value of: 36
  n=6
  7018 combinations were tried:
    The best set of stamps was 0 1 4 6 14 17 29
    With a maximum value of: 52
  n=7
  97394 combinations were tried:
    The best set of stamps was 0 1 4 5 15 18 27 34
    With a maximum value of: 70
  n=8
  1561926 combinations were tried:
    The best set of stamps was 0 1 3 6 10 24 26 39 41
    With a maximum value of: 93
  n=9
  29838545 combinations were tried:
    The best set of stamps was 0 1 3 8 9 14 32 36 51 53
    With a maximum value of: 121
```

Index